NH3 Energy+ 2018

AA008968

Topical Conference at the 2018 AIChE Annual Meeting

Pittsburgh, Pennsylvania, USA
28 October - 2 November 2018

ISBN: 978-1-5108-7636-1

Printed from e-media with permission by:

Curran Associates, Inc.
57 Morehouse Lane
Red Hook, NY 12571

Some format issues inherent in the e-media version may also appear in this print version.

Printed by Curran Associates, Inc. (2019)

For permission requests, please contact AIChE
at the address below.

AIChE
120 Wall Street, FL 23
New York, NY 10005-4020

Phone: (800) 242-4363
Fax: (203) 775-5177

www.aiche.org

Additional copies of this publication are available from:

Curran Associates, Inc.
57 Morehouse Lane
Red Hook, NY 12571 USA
Phone: 845-758-0400
Fax: 845-758-2633
Email: curran@proceedings.com
Web: www.proceedings.com

TABLE OF CONTENTS

Some abstracts in this book refer to images that were not made available by the publisher

540429 Power-to-Ammonia-to-Power (P2A2P) for Local Electricity Storage in 2025

Abstract:

A carbon-free, circular economy is required to decrease the greenhouse gases emissions. A commonly named alternative to the carbon-based economy is the *hydrogen economy*. However, storing and transporting hydrogen is difficult. Therefore, the *ammonia economy* is proposed [1–3]. Ammonia (NH_3) is a carbon-free hydrogen carrier, which can mediate the hydrogen economy [2, 4]. Especially for long-term storage (above 1 day), ammonia is more economically stored than hydrogen [5].

Transportation costs are greatly reduced by adopting a decentralized energy economy [6]. Furthermore, political-economic factors influence energy prices less in a decentralized energy economy. With small-scale ammonia production gaining momentum, business models for the decentralized ammonia economy are currently under development [6].

Within the current research, current technological advances are reviewed, as well as their feasibility for long-term, industrial application. A process flow diagram was developed, based on the *conceptual process design* method by Douglas [7]. An example case of $3.06*10^7$ kWh y^{-1} electricity requirement is used. The solution constitutes of wind and solar power, combined with a battery for short-term storage (up to 1 day) and a Power-to-Ammonia-to-Power process for long-term storage (above 1 day).

Technology pushes within the process design are the development of new materials (the Battolyser for a combined battery function and hydrogen generation, $Ru/Ba\text{-}Ca(NH_2)_2$ for ammonia synthesis, and $CaCl_2/SiO_2$ for ammonia separation and storage), enabling the decentralized production of ammonia from renewables at a low pressure (16 bar) [8–11]. Furthermore, the development of ammonia-fueled solid oxide fuel cells (SOFCs) enables electricity generation from ammonia directly, rather than from a cracked hydrogen and nitrogen feed stream [12, 13]. This simplifies the process design significantly (i.e., no ammonia cracker is required), and temperature swings within the process are decreased.

The combined ammonia separation and storage by $CaCl_2/SiO_2$ (a form of process intensification) is heat integrated with the SOFC, giving an electrical round-trip efficiency of ~30% for P2A2P (Power-to-Ammonia-to-Power). For comparison, the electrical round-trip efficiency of the battery function is ~80% for the Battolyser [11].

The total capital investment of such a system ($3.06*10^7$ kWh y^{-1}) is about 30-40 M€, including wind and solar power, a battery for short-term storage, and the Power-to-Ammonia-to-Power process. By modular design, the installation time is decreased and the working efficiency increased [14]. Depending on the electricity price (in the range 0.10-0.20 € kWh^{-1}), the payback period varies from 10 to 30 years.

Ammonia synthesis has been termed the *bellwether reaction*, and technological advancements go hand in hand with new complications [15–17]. Therefore, a roadmap is presented with potential improvements on current research and technologies, with a focus on mechanistic understanding of the processes involved. In case the desired technologies are not feasible, energy consumption penalties are indicated for alternative technologies.

In the upcoming years, it becomes technologically feasible to perform decentralized electricity storage in the form of ammonia. Ammonia as an energy vector is currently widely researched [18, 19]. Various technologies are competing (and transmutable), making the implementation of an ammonia economy technologically feasible. Furthermore, technological developments in the upcoming years can provide even better solutions for the decentralized ammonia economy. References

1. Avery, W. H. (1988). A Role for Ammonia in the Hydrogen Economy. *International Journal of Hydrogen Energy*, *13*(12), 761–773. doi:10.1016/0360-3199(88)90037-7
2. Christensen, C. H., Johannessen, T., Sørensen, R. Z., & Nørskov, J. K. (2006). Towards an ammonia-mediated hydrogen economy? *Catalysis Today*, *111*(1–2), 140–144. doi:10.1016/j.cattod.2005.10.011
3. Bartels, J. R., & Pate, M. B. (2008). *A feasibility study of implementing an Ammonia Economy*. Des Moines (IA).
4. Schüth, F., Palkovits, R., Schlögl, R., & Su, D. S. (2012). Ammonia as a possible element in an energy infrastructure: catalysts for ammonia decomposition. *Energy Environ. Sci.*, *5*(4), 6278–6289. doi:10.1039/C2EE02865D
5. Vrijenhoef, H. (2017). Dutch initiatives to store sustainable energy in the form of ammonia. In *NH3 Fuel Conference*. Minneapolis (MN). Retrieved from https://nh3fuelassociation.org/2017/09/25/dutch-initiatives-to-store-sustainable-energy-in-the-form-of-ammonia/
6. Brown, T. (2018). Small-scale ammonia production is the next big thing. Retrieved May 11, 2018, from https://ammoniaindustry.com/small-scale-ammonia-production-is-the-next-big-thing/
7. Douglas, J. M. (1988). *Conceptual Design of Chemical Processes. Mcgraw-Hill International.*
8. Malmali, M., Le, G., Hendrickson, J., Prince, J., McCormick, A., & Cussler, E. (2018). Better Absorbents for Ammonia Separation. *ACS Sustainable Chemistry & Engineering*. doi:10.1021/acssuschemeng.7b04684
9. Kitano, M., Inoue, Y., Sasase, M., Kishida, K., Kobayashi, Y., Nishiyama, K., ... Hosono, H. (2018). Self-organized Ruthenium-Barium Core-Shell Nanoparticles on a Mesoporous Calcium Amide Matrix for Efficient Low-Temperature Ammonia Synthesis. *Angewandte Chemie - International Edition*, *57*(10), 2648–2652. doi:10.1002/ange.201712398
10. Palys, M., McCormick, A., & Daoutidis, P. (2017). Design optimization of a distributed Ammonia generation system. In *NH3 Fuel Conference*. Minneapolis (MN). Retrieved from https://nh3fuelassociation.org/2017/10/01/design-optimization-of-a-distributed-ammonia-generation-system/
11. Mulder, F. M., Weninger, B. M. H., Middelkoop, J., Ooms, F. G. B., & Schreuders, H. (2017). Efficient electricity storage with a battolyser, an integrated Ni–Fe battery and electrolyser. *Energy Environ. Sci.*, *10*(3), 756–764. doi:10.1039/C6EE02923J
12. Eguchi, K., Takahashi, Y., Yamasaki, H., Kubo, H., Okabe, A., Isomura, T., & Matsuo, T. (2017). Development of materials and systems for ammonia-fueled solid oxide fuel cells. In *NH3 Fuel Conference*. Minneapolis (MN). Retrieved from https://nh3fuelassociation.org/2017/09/27/development-of-materials-and-systems-for-ammonia-fueled-solid-oxide-fuel-cells/
13. Okanishi, T., Okura, K., Srifa, A., Muroyama, H., Matsui, T., Kishimoto, M., ... Eguchi, K. (2017). Comparative Study of Ammonia-fueled Solid Oxide Fuel Cell Systems. *Fuel Cells*, *17*(3), 383–390. doi:10.1002/fuce.201600165
14. Bâldea, M., Edgar, T. F., Stanley, B. L., & Kiss, A. A. (2017). Modular manufacturing processes: Status, challenges, and opportunities. *AIChE Journal*, *63*(10), 4262–4272. doi:10.1002/aic.15872
15. Boudart, M. (1994). Ammonia synthesis: The bellwether reaction in heterogeneous catalysis. *Topics in Catalysis*, *1*(3–4), 405–414. doi:10.1007/BF01492292
16. Hellman, A., Honkala, K., Dahl, S., Christensen, C. H., & Nørskov, J. K. (2013). Ammonia Synthesis: State of the Bellwether Reaction. In *Comprehensive Inorganic Chemistry (II)* (2nd ed.). Elsevier Ltd. doi:10.1016/B978-0-08-097774-4.00725-7

17. Vojvodic, A., James, A., Studt, F., Abild-pedersen, F., Suvra, T., Bligaard, T., & Nørskov, J. K. (2014). Exploring the limits : A low-pressure , low-temperature Haber – Bosch process. *Chemical Physics Letters*, *598*, 108–112. doi:10.1016/j.cplett.2014.03.003
18. Soloveichik, G. (2016). Ammonia for Energy Storage and Delivery. In *NH3 Fuel Conference*. Los Angeles (CA). Retrieved from https://nh3fuelassociation.org/2016/07/06/ammonia-for-energy-storage-and-delivery/
19. Soloveichik, G. (2017). Future of Ammonia production: Improvement of Haber-Bosch process or electrochemical synthesis? In *NH3 Fuel Conference*. Minneapolis (MN). Retrieved from https://nh3fuelassociation.org/2017/10/01/future-of-ammonia-production-improvement-of-haber-bosch-process-or-electrochemical-synthesis/

540494 Cost Evaluation Study on CO_2-Free Ammonia and Coal Co-Fired Power Generation Integrated with Cost of CCS

Abstract:

This study presents a cost estimation for electricity generated by CO2-free ammonia and coal co-firing. Regulation of CO2 emissions seems to be gaining pace due to the global warming issue so the introduction of CO2-free energy in power generation has become desirable. Ammonia is one of the potential energy carriers for power generation and development of ammonia combustion technology with low NOx emissions has been conducted in Japan. In order to investigate the feasibility of the introduction of CO2-free ammonia in Japan from both the technical and economic viewpoints, we estimated the ammonia supply chain cost from ammonia production integrated with CCS in the UAE and applied this to ammonia distribution for an ammonia-coal co-fired thermal power plant.

We have assumed that the timing of the commercialization of such an ammonia-coal fired thermal power plant is around 2025 and determined the premise of the supply chain study by referring to related technical reports and information from experts. We estimated ammonia production cost in UAE, ammonia supply cost to a power plant in Japan and the cost of electricity produced by ammonia-coal co-fired power generation. The ratio of ammonia co-firing is 20 %-ammonia on high heat value(HHV) basis. The price of natural gas in UAE, the price of thermal coal in Japan, and cost of CCS in UAE and Japan is determined in accordance with the prices under current market conditions. Although there is no stated value regarding the cost of CCS in UAE and Japan, we have set these values as a matter of practical convenience so that we take information from public reports and information from experts into consideration. Regarding the cost of CCS in UAE, we set two different price types, one is the cost which is for CO2 originated from natural gas as feedstock and the other is from natural gas as utility fuel. The former cost is named "Cost of CCS in process" and the latter is named "Cost of CCS in utility". The reason for the different cost attributes is that a facility for CO2 capture and separation is introduced in the conventional ammonia production process. "Cost of CCS in process" is lower than "Cost of CCS in utility" by the reduction of the cost of CO2 capture and separation. Taking the ammonia supply cost into consideration, the electricity cost produced by ammonia-coal co-firing was estimated and we compared the estimated cost with the cost of electricity produced by conventional coal-fired power generation. Based on the estimation, we assessed the potentiality of ammonia as a CO2-free fuel for power generation.

This study was supported by the Council for Science, Technology and Innovation (CSTI), Cross-ministerial Strategic Innovation Promotion Program (SIP), "Energy Carriers" (Funding agency: JST).

522670 Ammonia-Hydrogen Power for Combustion Engines

Abstract:

Ammonia blends can potentially become a breakthrough chemical for power generation, cooling storage and distribution of energy. Gas turbines and internal combustion engines are potential candidates for the use of the resource in an efficient way that will enable commissioning of combined cycles to power communities around Europe and around the world while serving as sources of heat and chemical storage. Therefore, development of these systems will bring to the market a safer, zero carbon fuel that can be used for multiple purposes, thus decentralizing power generation and increasing sustainability in the communities of the future whilst positioning the developing and manufacturing companies as global leaders of a new generation of energy devices.

This work summarises results from various analytical, numerical and experimental campaigns carried out at Cardiff University, Wales, UK. Each of these is related to combustion of ammonia blended with hydrogen at different concentrations and conditions. Initially, computational work used to determine laminar flame speed, combustion products and ignition delay time is described. Software such as Gaseq, CHEMKIN-PRO and CANTERA were used to find the best mechanism suitable for NH_3/H_2 blend combustion. Next, experimental and numerical campaigns are described on the work carried out on a generic swirl burner employing NH_3/H_2 blend (50:50 by volume) at different equivalence ratios. Results showed that the flame stabilised above equivalence ratios of Ø=0.43. It was also observed that hydrogen seemed to be burning first, up to equivalence ratios <Ø=0.52, point at which the flame became unstable. Therefore, the results threw evidence of the narrow stability for high hydrogenated ammonia blends under lean conditions, steering the research to rich fuel combustion conditions. Production of NO_X was also measured, showing the high influence of OH radicals in the formation of the contaminant. Further works are presented for numerical, experimental and analytical studies where higher equivalence ratios were analysed at various conditions (pressure, inlet temperature and humidity) to determine the "sweet" spot where ammonia/hydrogen can be burned with the lowest emissions and highest stability. It was found that equivalence ratios between 1.2 and 1.3 provide the best point of operation, with relatively high humidity needed to bring down NOx emissions and unburned ammonia, consequence of the reactivity of intermediate radicals and flame compactness. Analytical studies determined the efficiencies of gas turbines operating with these blends, showing values that ranged from ~9% to 28% depending on the employed humidity. It is emphasized that further work is needed to determine equations and equipment that better characterise these efficiency values, as comparison needs to be carried out using feasible equipment that can withstand ammonia/hydrogen combustion under humidified conditions, equipment that does not exist at the moment.

Finally, results from numerical analyses and experimental studies using internal combustion engines are presented. Ammonia and hydrogen blends were evaluated in a methane-run Spark Ignition engine. A reduced chemical kinetic model is used with the closest-to-methane blend to determine cycle characteristics via CHEMKIN-PRO. Experiments demonstrate the potential of using these blends with moderate emissions and power outputs >15kW. However, it was

also found that the use of ammonia, due to its low energy content, needs to be carefully assessed to ensure that components, feeding lines and flowrates are precisely calculated before starting operation of these systems. Nevertheless, the results are encouraging, opening further interest in the development of more advanced injection strategies for these blends.

Overall, the motivation of the present study is to develop a â€œHydrogen through Carbon-Free Ammonia Economyâ€.

540587 Ship Operation Using LPG and Ammonia As Fuel on Man B&w Dual Fuel ME-Lgip Engines

Abstract:

Ship operation using LPG and ammonia as fuel on MAN B&W dual fuel ME-LGIP engines

Abstract

LPG has been used as fuel in the car industry for many years and now, with Exmar and Statoil's orders for ocean-going ships fitted with the dual fuel ME-LGIP engine, LPG will be used on marine engines as well. The new engine series is currently being developed to match all types of bigger merchant ships. This order was made in consequence of the new 2020 0.5% sulphur fuel cap, but this step forward has not stopped the discussion and interest in lowering CO2, NOx, SOx and particulate emissions even further. On the contrary, it has actually been further fuelled by the latest IMO meeting targeting a 50%-cut in greenhouse gas emissions from ocean-going shipping by 2050 compared with 2008.

Because the world fleet has increased since 2008, and thus CO2 emissions as well, it has been realised that this goal cannot be met without the use of carbon-free fuels. In shipping, 30 years corresponds to the lifetime of a ship, and owners will therefore soon need to consider this when they select the propulsion solution for their next ship. And as marine engine maker, MAN Diesel & Turbo needs to be fully prepared.

Using LPG as fuel on the two-stroke ME-LGIP engine offers similar emission benefits as with LNG, which reduces emissions significantly compared with MDO. Therefore, there are very good environmental reasons for using this fuel in coastal areas, on inland waterways and on deep sea. The LGIP engine solution system can also be applied on engine sizes from 5 to 85 MW, which are suitable for tankers, bulk carriers, container vessels, etc.

It is expected that the need to reduce CO_2 emissions will fuel a continued growth in shipping. And because sea transportation has proven to be less CO_2 polluting than both trucks and trains using fossil fuels this trend is expected to continue. Furthermore, the world population is increasing as well, and this is expected to increase the shipping fleet. So significant CO_2 reduction is mandatory in shipping, and this can be improved by using carbon-free fuels such as bio-LPG/LNG, and the so-called "electric fuels", etc. There are also plans to remove CO_2 from methane to produce carbon-free ammonia, but in order to be fully carbon free, the CO2 should be removed by, for example, injecting it into the seabed. MAN Diesel & Turbo already have dual fuel engines in our engine programme that can operate on LNG and methanol, but ammonia as a fuel has not yet been investigated for use on two-stroke engines.

This paper describes the technology behind the ME-LGIP dual fuel MAN B&W two-stroke engines, using LPG as fuel, and its associated LPG tank and fuel supply systems. The engine requires a gas supply pressure of 50-bar and controlled to a temperature of 45°C. At this temperature and pressure, the LPG is liquid, and different fuel supply solutions are available for generating this pressure for the liquid. Hence, the ME-LGIP for LPG will use liquid gas for injection, contrary to the ME-GI for LNG, where the methane is injected in gaseous form. All the way from tank to engine, the LPG remain in liquid phase, and conventional pumps can be used to generate the pressure. Furthermore, we have lately found that this engine technology, with minor modification, can also be used to burn ammonia, so the paper will also describe the modification needed in order to build an engine that is able to burn LPG as well as ammonia.

Safety is a concern when both LPG and ammonia is used as fuel on an engine located in an engine room. This is because LPG in gaseous form, contrary to methane, is heavier than air and will drop in case of leakage, and because ammonia in a gaseous form is toxic. This safety has been analysed, and our safety considerations and precautions will be described in details.

540742 Roadmap to All Electric Ammonia Plants

Abstract:

Haldor Topsøe A/S is a world leading supplier of technology and catalyst for the ammonia industry.

It is also a developer of Solid Oxide Electrolyzer technology. A road map towards all electrical ammonia plants of the future has been worked out implementing at first steps hybrid natural gas based/classical electrolyzer technology and ultimately SOEC based plants without air separation units.

540356 Ammonia As a Hydrogen Carrier for PEM Fuel Cells

Abstract:

Ammonia (NH_3) is easily liquefied by compression at 1 MPa and 25°C, and has highest volumetric hydrogen density of 10.7 kg H_2 /100L. It has high gravimetric hydrogen density of 17.8 wt%. The heat of formation of NH_3 is about 1/10 of combustion heat of hydrogen. NH_3 has advantages as a hydrogen carrier for fuel cell vehicles (FCVs).

ISO 14687-2:2012 specifies the quality characteristics of hydrogen fuel. The maximum concentration of NH_3 and N_2 for the FCVs is 0.1ppm and 100 ppm, respectively. The minimum H_2 purity is 99.97%. We need component technologies to produce high-purity hydrogen from ammonia, together with those to improve hydrogen energy efficiency.

High-performance ruthenium supported on MgO cracking catalyst was prepared. For this catalyst, NH_3 conversion was 99.8% at 0.1 MPa and 500°C which was almost the same as the chemical equilibrium value. We found that the remained NH_3 concentration of 1000 ppm was reduced to below 0.02 ppm using the remover including zeolite packed column. The NH_3 elimination quantity was 30-40 gNH_3/L. The Zeolite adsorbed NH_3 was recycled by annealing below 400°C. Hydrogen purification process with pressure swing adsorption (PSA) method was established to produce high purity hydrogen above 99.98% and to reduce nitrogen below 10 ppm. Micro-channel cracker with heat supply unit and H_2 purifier with off-gas supply unit were produced experimentally to improve efficiency.

In conclusion, we have developed component technologies to produce 1Nm^3/h-scale high-purity hydrogen from ammonia (NH_3 <0.02ppm, N_2 <10ppm, H_2 >99.98%) and to improve hydrogen energy efficiency by combination of the NH_3 cracker, the remover and the H_2 purifier. Hydrogen purification efficiency (hydrogen recovery rate) using the H_2 purifier was 90%. Hydrogen energy efficiency using the micro-channel cracker was 80 %.

I am greatly indebted to Professor T. Ichikawa and Associate Professor H. Miyaoka of Hiroshima University, Dr. T. Fujitani of AIST, Mr. T. Kuriyama of Showa Denko K.K., Mr. T. Adachi of Taiyo Nippon Sanso Corporation and Mr. H. Kubo of Toyota Industries Corporation for collaboration of this work. This work was supported by Council for Science, Technology and Innovation (CSTI), Cross-ministerial Strategic Innovation Promotion Program (SIP), â €œenergy carrierâ€ (funding agency : JST).

540631 Catalytic Membrane Reactors for Efficient Delivery of High Purity Hydrogen from Ammonia Decomposition

Abstract:

Abstract: The deployment of fuel cell electric vehicles is constrained by the paucity of hydrogen fueling stations and price, which is dominated by the costs of hydrogen storage and transportation. With more hydrogen per volume than liquid H_2 and an extensive distribution infrastructure in place, ammonia is a promising vector for efficient hydrogen distribution. In this talk we describe the development of innovative catalytic membrane reactor (CMR) technology for the delivery of high purity H_2 from ammonia cracking. The CMR integrates state-of-the art catalysts with thin metal membranes in an innovative design. Conventionally, the catalyst is supplied to CMRs in the form of a packed bed either in the interior of the support or in the annulus surrounding the membrane. In our design a dense palladium membrane a few microns in thickness is formed by electroless deposition on an asymmetric ceramic support. The catalyst is impregnated within the mesoporous layer adjacent to the membrane, and excellent dispersion maximizes its activity. The use of alkali promoters is found to further enhance the kinetics. The intimate proximity of the catalyst to the hydrogen permeable membrane provides a number of advantages over conventional systems. Radial transport limitations are mitigated, dramatically enhancing H_2 recovery. Ammonia decomposition is inhibited by H_2, but the efficient removal of H_2 in the CMR enables temperature reductions by up to 200 °C and decreases catalyst loading requirements by a factor of 3. Similar improvements in volumetric productivity over leading reports are obtained, reducing footprint requirements. We have developed a detailed axisymmetric model that captures the reactor performance with high fidelity, and this is used for the analysis and design of these systems.

540645 Development of a Highly Efficient CO_x-Free Ammonia Dehydrogenation System for Fuel Cell Applications

Abstract:

The shortage of fossil fuels and emission of carbon dioxide to the environment have attracted an interest in discovering renewable energy as the next generation energy source. Owing to its intermittent and unpredictable nature, however, excess renewable energy needs to be stored and reused on demand. In the regard, hydrogen, which possess a high gravimetric energy density and carbon free combustion process, has been extensively researched as a promising renewable energy carrier. However, the distribution and storage of hydrogen still raise important challenges due to the low volumetric energy density of hydrogen for its wide utilization. Currently, gaseous hydrogen transportation by pipeline and batch transportation using liquefied/compressed hydrogen have been implemented, but are either not economical viable particularly for long distance transport or significantly energy-intensive. Therefore, a lot of attention recently has been paid to the liquid chemical hydrogen storage materials such as liquid ammonia, cycloalkanes and heterocycles, formic acid, and so forth.

Among the candidate, liquid ammonia is an excellent hydrogen carrier owing to its high gravimetric and volumetric hydrogen storage capacities and moderate condensation conditions compared to other chemical hydrogen storage materials. Furthermore, the production and distribution activities of ammonia are already well-established processes internationally. The liquid ammonia can be dehydrogenated at the temperatures of â‰¤ 550 ËšC leading to hydrogen and nitrogen as major products with no carbon release to the environment. All these properties make liquid ammonia highly intriguing as a hydrogen carrier for power generation in conjunction with fuel cells.

The present study demonstrates a > 1kW-class COx-free power generation system including an ammonia feed, a dehydrogenation reactor, a i-butane burner, a heat exchanger, a hydrogen purification unit and a PEMFC. To the best of our knowledge, no study has reported a NH_3/H_2 based sustainable energy conversion system in a practical scale. With respect to the catalyst, Ru supported on La-doped alumina (Ru/La(x)-Al_2O_3) with different lanthanum loadings were used in the form of pellets. The dehydrogenation activity of the catalyst was optimized at a 20 mol% lanthanum loading and a reaction temperature of 500 ËšC, reaching close to complete conversion of ammonia. Catalyst characterization results revealed close to 1.5 wt% Ru metal loading of the catalyst and 2-3 nm Ru nanoparticles. Moreover, various absorbent materials for hydrogen purification process were tested to remove unreacted ammonia from the product stream. Among the materials tested, 13X zeolite indicated the best performance acquiring ammonia concentration of 650 ppb after the absorbent tower.

Overall, the as-integrated system fueled by ammonia was able to hold 9 $Lmin^{-1}$ of NH_3 at 550 ËšC, producing 13.4 $Lmin^{-1}$ of hydrogen at an ammonia conversion of more than 99.7%. This amount of hydrogen was able to power 1kW-class PEMFC continuously for over 2 h without any loss in the fuel cell performance. The calculated reformer and system efficiencies were considerably improved by recirculating the remaining hydrogen from the exhaust PEMFC, which was then re-entered to the burner. In this configuration, we were able to achieve COx-free operation conditions (no i-butane was utilized) with the reformer and system efficiencies of 84.4% and 49.7%, respectively. Finally, a test drone modified using a commercial hexacopter was tethered to the as-developed hydrogen power pack fueled by ammonia where the electrical power was provided. The drone was able to maintain significantly longer flight time in contrast to the onboard battery operation (14 min, no payload). The estimated flight time was approximately 4.1 h using a 3.4 L NH_3 tank.

535457 Material Discovery and High Throughput Exploration of Ru Based Catalysts for Low Temperature Ammonia Decomposition

Abstract:

High throughput experimentation gives us the unique ability to generate massive, multidimensional datasets that are not typical for heterogeneous catalysis studies. Here, we show the synthesis and catalytic screening of over 100 different Ru based bimetallic catalyst combinations using 33 different metals that were synthesized via incipient wetness impregnation. The catalysts were analyzed using Wide Angle X-ray Scattering (WAXS) for phase identification. Catalysts were screened for ammonia decomposition activity using a 16-channel parallel plug flow reactor. Fourier transform infrared (FT-IR) imaging was used to analyze all 16 effluent streams in parallel in under one minute.

All results obtained from WAXS characterization and catalyst screening were fed into a machine learning algorithm to extract the activity descriptors and elemental characteristics that are responsible for ammonia decomposition activity at different operating temperatures. The knowledge extracted from this materials agnostic machine learning algorithm was used to design a second iteration of catalysts, where features that contributed to the greatest change in activity were accentuated.

519851 Functionalized Ordered Mesoporous Silica Composites As Potential Ammonia Storage Materials

Abstract:

Ammonia may provide an alternative energy supplier for its strong capability as hydrogen carrier. However, it is a problem that how to storage this kind of chemical at relatively high temperature, for example 300°C in fuel cell. In this work, a composite material based on metal halides and ordered mesoporous silica framework is developed and used to target ammonia at relatively high temperature. The silica framework is fabricated via evaporation induced self-assembly method and has tunable mesoporous structure with addition of hexadecyl trimethyl ammonium bromide (CTAB). Several metal salts at various concentrations are added to the mesoporous framework via wetness impregnation method. It is reported that ammonia storage capability is further enhanced by functionalization with relative metal sites. Along with capacity measurements, the incorporation of metal salts on mesoporous silica is studied using microscopy, X-ray diffraction and porosity characterization techniques. Results show that well-dispersed metal crystal trapped in submicron-sized porous structure can improve diffusion rate and increase pore accessibility while maintaining structural integrity at relatively high temperature. Capacity measurements show that functionalized metal/silica composite can work as potential materials for ammonia storage. It is also a thermal stable material for only 3% weight loss over 300°C.

540413 Development of Catalytic Reactors and Solid Oxide Fuel Cells Systems for Utilization of Ammonia

Abstract:

Hydrogen is the primary fuel source for fuel cells. However, the low volume density and difficulty in storage and transportation are major obstacles for the practical utilization. Among various hydrogen carries, ammonia is one of the promising candidates because of its high hydrogen density and boiling point and ease in liquefaction and transportation. The reaction temperature of ammonia cracking to nitrogen and hydrogen, being about 600°C or higher, is close to the operating temperature of solid oxide fuel cells (SOFCs). The integration of these two devices is beneficial in terms of heat and energy managements and will lead to the development of simplified power generation systems.

Several catalyst materials, i.e., cracking, autothermal cracking, and combustion of ammonia have been investigated and used for the SOFC systems. A demonstration of the stack-level ammonia-fueled SOFC systems is an important step for the actual utilization of ammonia-fueled SOFCs. In this study, 200 W class and 1 kW class SOFC stacks were applied for ammonia fueled generation systems.

Catalytic decomposition of ammonia

Catalysts for decomposition of ammonia has been developed. The activity of Ni/Y_2O_3 and SrO modified Ni/Y_2O_3 for ammonia cracking was sufficiently high for combination with SOFCs. The SrO modified Ni/Y_2O_3 catalyst showed higher performance than Ni/Y_2O_3 and achieved complete decomposition at 600ºC. The ammonia decomposition completed at lower temperatures than that of the typical operation of SOFC (750ºC). This means the ammonia cracking can proceed without an external heating in the integrated system with SOFCs. This catalyst maintained high activity during the operation for 1000 h at 700ºC.

Autothermal decomposition reactor

Catalyst based on Co-Ce-Zr composite oxide has been developed for autothermal ammonia cracking. Autothermal cracking of ammonia is characterized by fast start-up in the exothermic condition. It is necessary to operate the autothermal ammonia cracker at a suitable air to fuel ratio (AFR) condition, considering the electrical efficiency as well as the amount of heat generated from the cracker. The AFR value for the supply of the autothermally cracked ammonia to the SOFC stack was set to be 0.75, since a sufficient heating value and a high ammonia conversion could be expected. The temperature increased at the initial stage with a honeycomb shaped metal heater until the inlet temperature of the catalyst honeycomb reached at 200ºC. Then, the temperature continuously rose up without electrical heating with the exothermic reaction in the catalyst honeycomb. The start-up time required from the initiation of electrical heating to the achievement of the steady state was 130 s. Therefore, it is concluded that this reactor is suitable for start-up and heating of the SOFC system.

SOFC stack fueled with ammonia

The ammonia fueled SOFC stack systems were evaluated. The cell consisted of a Ni/ZrO_2-based fuel electrode, ZrO_2-based electrolyte, and perovskite-type oxide air electrode. The planar single cell was 120 mm in diameter. Each cell was connected to metallic separator plate with internal manifolds and gas tightness was ensured with a glass sealant. The stack composed of 10 single cells was evaluated from its *I–V* and *I–P* characteristics at 770ºC. Four fuel supply systems were connected to the stack, i.e., 1) direct supply of dry ammonia, 2) the ammonia cracker equipped with the Ni/Y_2O_3 catalyst operated at 600ºC, 3) the autothermal ammonia cracker equipped with the catalyst honeycomb operated at an AFR of 0.75, 4) The mixture of hydrogen and nitrogen with a composition of 3 to 1 for comparison. The open circuit voltages of the direct ammonia and cracked ammonia fueled SOFC stacks were almost the same as that of the hydrogen fueled stack. This means, for the direct ammonia fueled stack, the supplied ammonia was almost completely decomposed into hydrogen and nitrogen over the anode. In the case of autothermally cracked ammonia, the open circuit voltage of stack was slightly lower than the other fuel supply systems because of steam in the supply gas. The electrical power achieved with this 10-cell stack was about 250 W. The performances for the SOFC stack was comparable for four fuel supply systems.

One kW-class SOFC stacks and packaged system

The stack consisting of 30 single cells was tested at *ca.* 750ºC. Since the performance of the cells has been improved, the power per cell in this stack has also been improved as compared with the cells used in 10-cell stack. The power of 1073 W can be achieved with the supply of dry ammonia at the current of 50 A which was almost the same as the supply of $3H_2+N_2$ mixture at the same current. The average cell voltage at this operation point was 0.715 V. The DC efficiency was 52.3% at the fuel utilization of 80%. The 1 kW class stack was successfully operated for 1000 h. The 1kW-class stack was also operated by supplying the reacted gas from autothermal ammonia decomposition reactor.

The packaged SOFC system with automatic start-up and shut-down sequences was tested. Thermally self-sustainabile 1 kW system was successfully operated.

Development of catalysts for ammonia combustion

For heating of the system and cleaning of ammonia containing exhaust, ammonia combustion catalysts have to be developed. The activity for catalytic combustion of ammonia was higher over supported Pd catalysts than over Pt or other precious metal catalysts. It is important for combustion catalyst to achieve high selectivity to N_2 without producing N_2O, NO, NO_2, and other NO_x species. On most of the supported Pd catalysts, combustion of ammonia started from 200ºC followed by steep enhancement of conversion. The selectivity to N_2 and NO_x species depended strongly on the support oxide. It was found that Pd supported on zeolite is effective in combustion of ammonia without emitting NO_x species and N_2O.

This work was supported by the Council for Science, Technology and Innovation (CSTI) Cross-ministerial Strategic Innovation Promotion Program (SIP) "energy carrier" (Funding agency: JST).

Importance of Reaction Mechanism Involved in Design of the Catalyst and the Reactor for Future Ammonia Synthesis

Presentation ID# 540411

Dr. Ken-ichi Aika, Tokyo Institute of Technology, Meguro-ku, Tokyo, Japan

Abstract

The ammonia synthesis reaction is considered to involve several elementary steps [1]:

$N_2 + 2*$ à $2N(a)$ (1)

$H_2 + 2*$ à $2H(a)$ (2)

$N(a) + H(a)$ à $NH(a) + *$ (3)

$NH(a) + H(a)$ à $NH_2(a) + *$ (4)

$NH_2(a) + H(a)$ à $NH_3(a) + *$ (5)

$NH_3(a)$ à $NH_3 + *$ (6)

Here, the symbol * indicates empty sites. For most metal catalysts, the dissociative adsorption of dinitrogen (step 1) is the rate-determining step, and all the other steps and its reverse step (from 2 to 6) are fast enough to be almost in equilibrium for each reaction step. The synthesis rate is considered to be the same as the rate for step (1), and is proportional to the dinitrogen pressure and square of the empty site concentration. For iron catalysts, the main adsorbed species are simply N atoms, and the empty site concentration is given by $[*] = 1-\theta_N$, where θ_N is the surface coverage of N. Under equilibrium conditions, Eqs. (2) to (6) can be combined to give the following equation:

$N(a) + 1.5H_2 = NH_3 + *$ (7)

The equilibrium constant in Eq. (7), $1/K_d$, is expressed as:

$1/K_d = (1-\theta_N)P_{NH3}/\theta_N P_{H2}^{1.5}$ (8)

The empty site concentration, $(1-\theta_N)$, is given by:

$(1-\theta_N) = 1 / (1+ K_d\, P_{NH3}/ P_{H2}^{1.5})$ (9)

Thus, the synthesis rate can be expressed as a function of the N_2, H_2, and NH_3 pressure, as follows:

$r_1 = k_1 P_{N2} (1-\theta_N)^2 = k_1 P_{N2} (1+ K_d\, P_{NH3}/ P_{H2}^{1.5})^{-2}$ (10)

Here, k_1 is the rate constant for step (1).For ruthenium catalysts, dissociative adsorption is the rate-determining step, as with other metals, but adsorbed hydrogen has been found to retard the synthesis reaction. When both N(a) and H(a) retard dinitrogen activation, rate equation (10) is expressed instead as follows:

$$r = k_1 P_{N2} (1 + K_d P_{NH3} / P_{H2}^{1.5} + K_2 P_{H2}^{0.5})^{-2} \quad (11)$$

Here, K_2 is the equilibrium constant for hydrogen adsorption (step (2)).

Thus character of the catalyst is expressed with the values of k_1 (ability of N_2 activation), K_d (strength of N adsorption), and K_2 (strength of H adsorption). Active ammonia catalyst has high value of k_1. Generally K_d is great at iron but low at ruthenium. On the other hand K_2 is great at some ruthenium catalyst (strong hydrogen inhibition. The surface nature of ruthenium catalyst is strongly affected with the support and promoter. This is the reason why a lot of unique catalysts have been reported recently. The relation between promoter compounds and kinetic indexes shall be shown in this presentation. These indexes are necessary for the reactor design and process operation (pressure, temperature, conversion, H/N ratio). If we can install different kind of catalysts, we can set a catalyst with great K_d at the reactor inlet part, but not at the outlet part where produced ammonia depress the rate by strong K_d. Such examples shall be shown.

Another interesting issue is the use of effective ammonia separation material for future process. We have proposed how to design proper separation materials by mixing alkaline earth halides [2]. These problems shall be also discussed in this presentation. Future ammonia production condition might be changed when renewable hydrogen is used. This kind of study must become important.

[1]; K. Aika, "Role of alkali promoter in ammonia synthesis over ruthenium catalysts—effect on reaction mechanism", Catal. Today, 286 (2017) 14-20.

[2]; C. Y. Liu, K. Aika, Ind. Eng. Chem. Res., 43 [22], 6994-7000 (2004); ibid. 43 [23], 7484-7491 (2004).

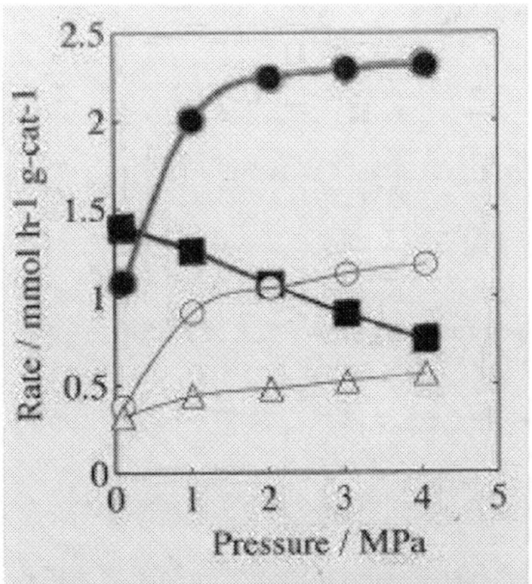

Fig.1 NH_3 sythesis rate as a function of total pressure ($N_2 + 3H_2$) at 588K over 3wt% Ru-Cs^+/MgO(■), 3wt% Ru/CeO_2-LTR(○), Ru/CeO_2-HTR(●), and 3wt%Ru/MgO_2(△).

Low H_2 inhibition on Ru/CeO_2

SMSI: Strong Metal Support Interaction (support oxide partly reduced)

Y. Niwa and K. Aika, J. Catal., **162, 138-142 (1996).**

AIChE 2018 Oct 31

2

Proposed mechanism of ammonia synthesis on Ru/CeO2 (reduced at high temp) and Ru-CsOH/MgO

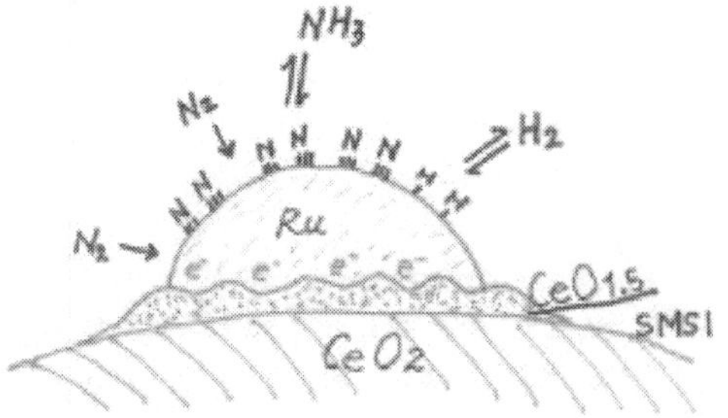

CeO2 is partly reduced (SMSI) and delivers electron to Ru

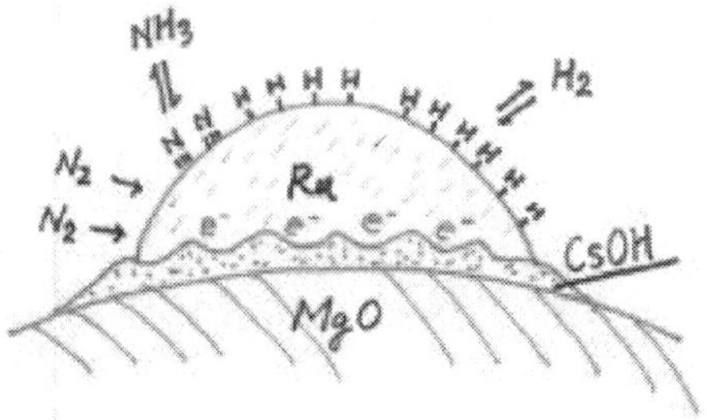

CsOH sticks to Ru and delivers electron to Ru

Mechanism and the rate of catalytic ammonia synthesis

$N_2 + 2* \rightarrow 2N(a)$ (1) rate determining step

$H_2 + 2* \rightarrow 2H(a)$ (2)

$N(a) + H(a) \rightarrow NH(a) + *$ (3)

$NH(a) + H(a) \rightarrow NH_2(a) + *$ (4)

$NH_2(a) + H(a) \rightarrow NH_3(a) + *$ (5)

$NH_3(a) \rightarrow NH_3 + *$ (6)

$N(a) + 1.5H_2 = NH_3 + *$ (7)

The equilibrium constant in Eq. (7), $1/K_d$, is expressed as:

$1/K_d = (1-\theta_N)P_{NH3}/\theta_N P_{H2}^{1.5}$ (8)

$(1-\theta_N) = 1/(1+K_d P_{NH3}/P_{H2}^{1.5})$ (9)

Thus, the synthesis rate can be expressed as a function of the N_2, H_2, and NH_3 pressure, as follows:

$r_1 = k_1 P_{N2}(1-\theta_N)^2 = k_1 P_{N2}(1+K_d P_{NH3}/P_{H2}^{1.5})^{-2}$ (10) for Fe catalysts

$r = k_1 P_{N2}(1+K_d P_{NH3}/P_{H2}^{1.5} + K_2 P_{H2}^{0.5})^{-2}$ (11) for Ru catalysts

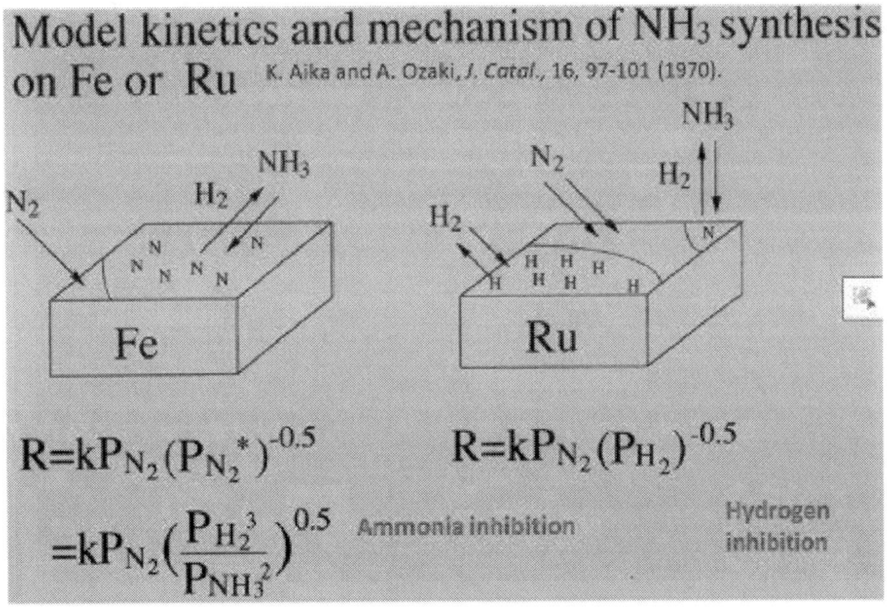

AIChE 2018 Oct 31

5

$N_2 + 3H_2 \rightarrow 2NH_3$

($H_2/N_2=3$) gas is supplied as a whole, but we can control the reactor inlet concentration with ($H_2/N_2=1$).

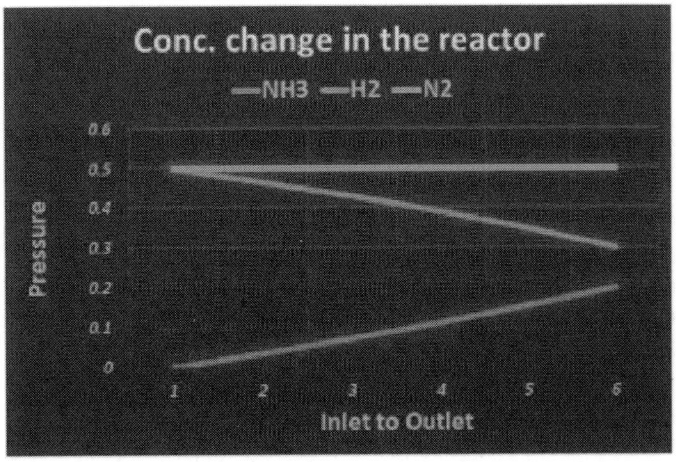

H2 conv. 50%

Gas conv. 20%

Catalyst setting → Ru/CeO2 | Ru-CsOH/MgO →

AIChE 2018 Oct 31

6

NH3 synthesis from renewable energy: proposed unit

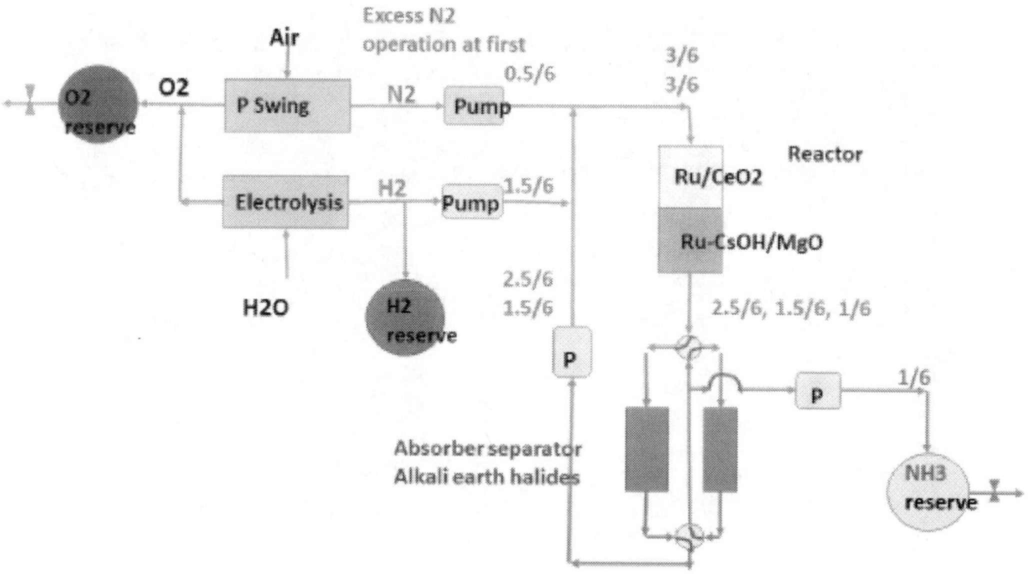

Ammonia reversible absorption to $CaCl_{2-x}Br_x$ (Liu, Aika; 2002)

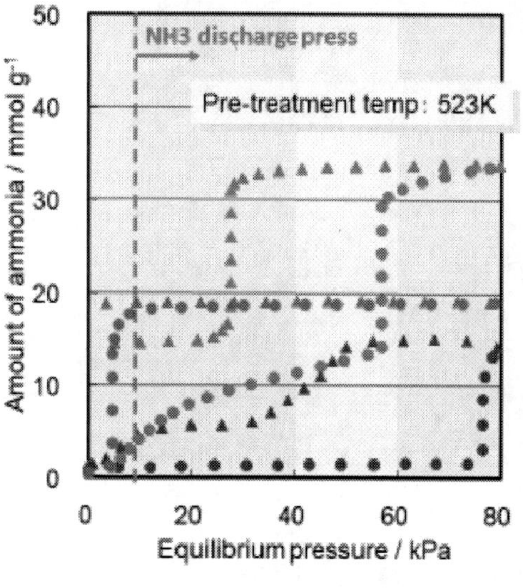

Reversible absorption temp: 298K
●: Absorption isotherm
▲: Desorption isotherm

$CaCl_2$, CaClBr, $CaBr_2$

Anion mixed Ca halide shows intermediate character of those pure compounds

C. Y. Liu, K. Aika, Ind. Eng. Chem. Res., 43 [22], 6994-7000 (2004); ibid. 43 [23], 7484-7491 (2004).

Reversible ammonia absorption on $CaCl_{2-x}Br_x$ (Liu, Aika 2002)

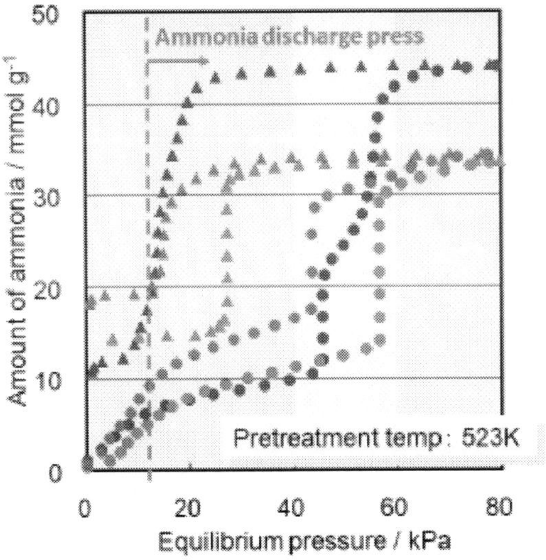

NH3 reversible absorption temp: 298K
●: Absorption isotherm
▲: Desorption isotherm
$CaCl_{1.33}Br_{0.67}$, CaClBr, $CaCl_{0.67}Br_{1.33}$

Absorbs much amount of NH3 below 60kPa, desorbs NH3 above 10kPa reversibly.

⇒ Pressure swing separation of NH3 is possible under mild condition.

$CaCl_{2-x}Br_x$ ammine complex formation model

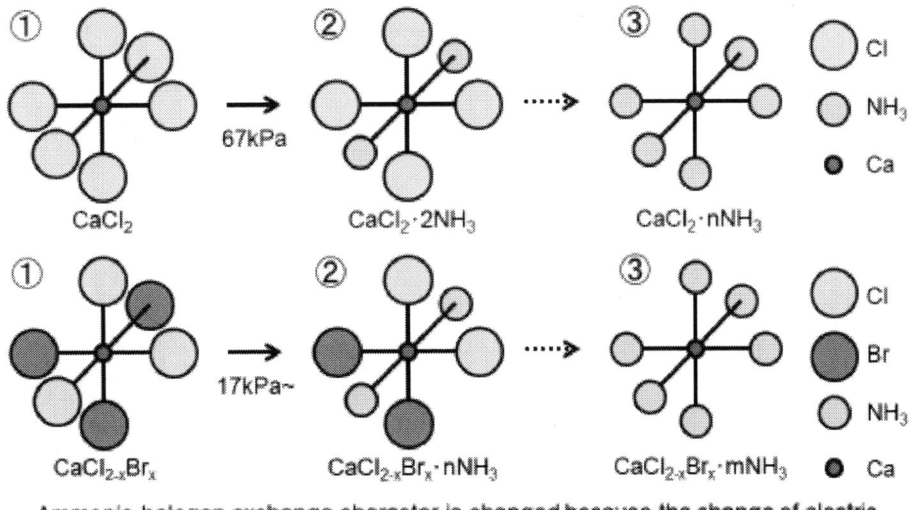

Ammonia-halogen exchange character is changed because the change of electric field and/or structure around Ca ions. > Controls NH3 absorption pressure.

531040 Ammonia Absorption and Desorption in Ammines

Abstract:

Ammonia Absorption and Desorption in Ammines

Collin Smith, Mahdi Malmali, Chen-Yu Liu, Alon McCormick, and E L Cussler

University of Minnesota, Minneapolis MN 55455

(abstract)

While adsorption onto solids is a common separation process, absorption into solids is much less often used. The reason is that absorption is usually assumed ineffective because it includes very slow solute diffusion into the solid. An exception may be the separation of ammonia from nitrogen and hydrogen using ammines, especially at temperatures close to those used in ammonia synthesis. There, ammonia can be selectively absorbed by calcium chloride; nitrogen and hydrogen are not absorbed. The kinetics of ammonia release seem to be diffusion controlled. The kinetics of absorption are consistent with a first order reaction and diffusion in series, so the rate controlling step changes with the amount of absorption. The potential of this separation in the distributed production of ammonia is discussed both as a source of fertilizer and as a means of wind energy storage.

525808 Scale up and Scale Down Issues of Renewable Ammonia Plants: Towards Modular Design

Abstract:

Renewable sources of energy such as biomass, solar, wind or geothermal just to mention some of the most widely extended are characterized by a highly distributed production across regions (EPA, 2017). Total renewable energy available is more than enough to provide for society needs, but the traditional production paradigm is changing. Economies of scale have featured current industry and its infrastructures based on large production complexes (i.e Dow, Exxonmobil or BASF hubs). The well-known six tenths rule has extensively been used in the chemical industry to scale up or down the cost of technologies. This rule is suitable for large chemical complexes, where economies of scale are a critical point. However, distributed production completely changes this paradigm. Distributed production corresponds to the production at small scales (Pepermans et al., 2005). This new production scheme results in the use of a number of individual units so that the cost is no longer a continuous function. The step forward is modularization of chemical plants. The plants will be built in the form of modulus that are easily assembled at any place to make the most of distributed resources (Baldea, 2017). The advantages of these plants are straightforward, easy and quick deployment and low investment risk to exploit resources even in remote places. However, the disadvantages must be also considered including the environmental impact of the transportation of the plants and of the products (EPA, 2017).

Large ammonia plants based on fossil resources dominate industry (Appl, 2011). It is a power demanding industry. However, nitrogen and hydrogen can be produced from renewable sources (Sánchez and Martín, 2018). In this work, a process level analysis for the optimal use of distributed sources of energy at various scales applied to the production of ammonia is performed. The scale up and down of facilities is evaluated selecting the proper technology for each of the three sections of the plant, evaluating the technologies and their economics. The facility consists of four stages, power collection, considering wind turbines or solar PV panels, air separation for which membranes, PSA or distillation technologies are evaluated, water electrolysis and ammonia synthesis evaluating two reactor designs. The process is optimized by solving a MINLP for several production capacities evaluating the operating and investment costs resulting from the need to use a number of parallel units membranes and PSA systems.

The production capacity defines the best technology depending on its characteristics. A number of trade-offs are seen in the results. While membranes are suitable for very small production capacities, PSA system for medium and distillation column for large, PSA systems are competitive across scales, but the large number of units required can be a drawback to pursue this technology that requires lower operating pressure than distillation columns. Modular designs are only efficient and competitive, in economic terms, when full capacity is used, however this design presents other benefits have been not evaluated, for example, are versatile in their operation. The large energy consumption has larger impact in the investment of large facilities, where energy collecting units represent 60% of the production costs. However, smaller capacities investment cost is governed by chemical units. Finally, the expected decrease in PV panels is far larger than that of wind turbines, representing 90% savings in PV panels and very promising production costs for ammonia below 1€/kg. These saving can reduce the investment cost by half in the near future.

References

Environmental Protection Agency (EPA), 2017. Distributed generation of electricity and its environmental impacts https://www.epa.gov/energy/distributed-generation-electricity-and-its-environmental-impacts Last accessed: February, 2018.

Pepermans, G., Driese, J., Haeseldonckx, D., Belmans, R., D'haeseleer, W.D., 2005. Distributed generation: Definition, benefits and issues. Energy Policy, 33 (6), 787-798.

Baldea, M., Edgar, T.F., Stanley, B.L., Kiss, A.A., 2017. Modular manufacturing processes. Status, challenges and opportunities. AIChE J., 63 (10), 4262-4272.

Sánchez, A., Martín, M., 2018. Optimal renewable production of ammonia from water and air. J. Clean. Prod., 178, 325-342.

Appl M (2011) Ammonia 2.- Production processes. Ullmann's Encyclopedia of industrial chemistry. DOI: 10.1002/14356007.o02_o11

540509 Advances in Making High Purity Nitrogen for Small Scale Ammonia Generation

Abstract:

The presentation will address recent developments in the Solar Hydrogen Demonstration Project in which hydrogen, nitrogen and ammonia are made from solar power, water, and air; and used to fuel a modified John Deere farm tractor.

In industrial applications very pure nitrogen is made by cryogenic distillation of air. Using Pressure Swing Absorption systems alone it is extremely difficult to achieve the required purity. An improved method was developed for making high purity nitrogen, for smaller systems.

Will discuss how, when Oxygen contaminates the reactor catalyst, Hydrogen is used to purge the catalyst, and subsequently used as fuel.

532661 Vanadium As a Potential Catalytic Membrane Reactor Material for NH_3 Production

Abstract:

In solid or liquid states, ammonia salts and solutions are the active components of most synthetic fertilizers used in agriculture, which consume 83% of the worldâ€™s ammonia. Today, ammonia for fertilizers is industrially produced via the Haber-Bosch process at 400-500 Â°C and at pressures up to 30 MPa (300 bar). These harsh operating conditions are necessary due to the high affinity of dissociated nitrogen atoms towards the catalyst surface in addition to the high barrier associated with N_2 dissociation. For these reasons, the need for advanced catalytic methods for the reduction of N_2 to ammonia remains a requirement for sustainability in the food and energy cycle.

The aim of this work is to explore the potential of metallic membranes for N_2 separation with the final intent to produce NH_3. Based on a preliminary theoretical investigation using density functional theory, the Group V transition metals (e.g., vanadium (V), niobium (Nb) and tantalum (Ta)) show strong affinity toward N_2. Moreover, from solubility and diffusivity values taken from the literature, iron (Fe) is a suitable fit for this application. The first experimental study showed that V, Ta, Nb, and Fe have a N_2-permeating fluxes on the order of 10^{-5} and 10^{-4} $mol_{N2}/m^2 \cdot sec$, depending on the operating condition used and that V-based metallic membranes have excellent properties for the preferential transport of N_2 over CO_2 and CH_4 with near-infinite selectivity

The V metal is, hence, used as a potential membrane to produce NH_3. Specifically, it is housed in the membrane reactor where a H_2 stream is used as the sweep gas to promote the NH_3 reaction. Lower pressures than the conventional Haber-Bosch process are used. Specifically, a range of operating pressure from 30 to 60 bar, instead of 200 bar (Haber-Bosch process), and a range of temperature from 350 °C to 500 °C are used as conditions to produce NH_3. The performance of the membrane reactor in terms of NH_3 conversion is, hence, investigated.

540706 Advanced Catalysts Development for Small, Distributed, Clean Haber-Bosch Reactors

Abstract:

The traditional Haber-Bosch (HB) synthesis of anhydrous ammonia will adapt to clean power by sourcing the hydrogen from renewable electrolysis. However, the very large scale of current HB plant designs are not well-matched to smaller and more distributed clean power resources. Plant/reactor designs need to be made at a smaller scale in order to best utilize clean hydrogen. Small, megawatt scale HB reactors have an additional advantage of being better able ramp up and down with variable renewable power. This talk will detail ARPA-e funded work into the design and optimization of these smaller, clean NH3 reactors, which utilize much higher and variable space velocities, lower pressures, and gas adsorption rather than condensation for NH3 extraction.

The different synthesis conditions at smaller scale also require a rethinking of the existing HB iron catalysts, which have been optimized for large, low space velocity reactors. New, advanced heterogeneous catalysts, capable of 5X faster nitrogen fixation, will be discussed. This talk will detail the development of the catalyst, including the ternary oxide support and metal nanoclusters, both of which play a key role in the catalyst performance. Many phases of the ternary system were synthesized, characterized for surface area and metal dispersion, and tested in a differential reactor to map catalytic performance to support composition and structure. Additionally, different active metals were attached to the best performing oxide support, to study the activity dependence on metal species.

536332 Ammonia Synthesis Via Radiofrequency Plasma Catalysis

Abstract:

Introduction:

In 1909, a compound named Ammonia was discovered. Through the 20th century, the immense potential of this chemical was exploited by using in almost every product, from process industry for fertilizer and chemical production to every use in cosmetics, household cleaners and medicines. Recently, fuel cells operating on liquid ammonia as working fluid have been developed on research scale. Despite of using 1-2% of total energy production for the synthesis of this compound, no significant changes have been made to the process since the first Haber-Bosch process plant has been setup.

Plasma catalysis is the use of plasma and catalyst synergism for the synthesis of various compounds. In case of ammonia synthesis, it helps in shifting the rate-limiting step from nitrogen dissociation to NH_x formation. The excitation source helps the molecules to reach excited and ionized states which ensures the abundance of radicals for radicals. Radio-Frequency plasma is once such tool for plasma-catalysis. The synthesis has been explored by Matasumoto et al.[1] but no concrete details about the reaction pathway and plasma-catalyst synergism have been reported.

Materials and Methods:

In this study, we synthesize ammonia in an in-house built plasma reactor. Inductively coupled Radio-frequency has been used as the excitation source with a Quartz tube under vacuum. The synthesis is carried at 400 â„ƒ and the exhaust gases from the pump are used for analyses. The catalyst used were 1g meshes of Fe, Cu, Pd, Ag and Au. A semi-metal Ga, was used as a catalyst for the very first time for ammonia synthesis for any known synthesis process

Results:

Yields as high as 19% have been achieved at 300 W plasma power and Au catalyst. Although, the highest synthesis rate achieved is 0.22 g-NH_3/kWh at 50 W plasma power with Ga catalyst. The simulations revealed the rate-limiting step as NH_2 formation. The emission spectra revealed the dominance of gas phase atomic hydrogen ($H_{É}$·) species on ammonia yield. Based on the experimental results and simulation we elucidate the mechanism predominantly comprising of surface reaction on the wall of the reactor and the surface of catalyst.

[1] Tanaka, S., Uyama, H., & Matsumoto, O. (1994). Synergistic effects of catalysts and plasmas on the synthesis of ammonia and hydrazine. *Plasma Chemistry and Plasma Processing*, *14*(4), 491-504.

541115 Terrestrial Energy, National Lab, Southern Company â€“ Partnership Overview Using Integral Molten Salt Reactor Technology with Hys Acid for Hydrogen Production

Abstract:

Demands for safe secure supplies of potable water across the planet are increasing faster than can be provided by natural, ever depleting sources of fresh water. At the same time, world demand for electric power is also accelerating.

Making H_2 from Natural Gas is not an optimal or efficient process that is also un-economic at higher gas costs. An Integral Molten Salt Reactor (IMSR) is uniquely suited to provide the very high temperatures (585°C+ working temps.) that are needed to both generate significant amounts of Hydrogen, Oxygen (a feed for industrial oxygen uses) and Electricity needed for advanced economies and industry.

Terrestrial Energy USA, Savanna River National Lab, Sandia National Lab, Idaho National Laboratory, and Southern Company are working to show that this new generation of IMSR would be the only effective system to create an H_2 supply utilizing the HyS system to produce Hydrogen.

The Hybrid Sulfur (HyS) Hydrogen Generation process has the potential to produce hydrogen gas using both thermal and electrical energy at a cost of <$2/kg.

HyS can utilize thermal energy from a molten salt reactor (MSR) along with renewable electrical energy from either wind or solar generation to efficiently produce hydrogen.

The HyS process, being a two step process, can act as a buffer and store thermal energy chemically as liquid SO_2, to be used to generate hydrogen as required to minimize generation and storage costs.

Analysis by INL and Terrestrial USA have shown that IMSR is uniquely suited to be coupled to a pure H_2 production from HyS.

Deep de-carbonization of Process Heat, H_2 production and power demands could both be served by a carbon free source of inherently safe, clean, non-proliferating energy.

Other findings of the work studies show that in addition to performing very well with H_2 leading to Ammonia production, there are many other IMSR H_2, O_2 / Energy production combinations available that can be tailored to each application and industry need. The ability to immediately dispatch heat energy to where it is needed is a specific trait of the IMSR.

With an unequaled, walk away safe source of compact, efficient, very high temperature process heat, coupled with the latest industrial systems, Terrestrial Energy USA, working with United States Industry and National Labs, can be the catalyst for an vastly expanded NH_3 Industry with the H_2, O_2 and power needed to make an industrial economy flourish.

529953 Creating a Redox Materials Database for Solar-Thermochemical Air Separation and Fuels Production

Abstract:

Creating a Redox Materials Database for Solar-Thermochemical Air Separation and Fuels Production

J. Vieten[a], P. Huck[b], D. Guban,[a] M. Horton[b], B. Bulfin[c], M. Roeb[a], K.A. Persson[b], C. Sattler[a]

[a] Institute of Solar Research, German Aerospace Centre (DLR), Cologne, Germany

E-mail address: Josua.Vieten@dlr.de

[b] Environmental Energy Technologies Division, Lawrence Berkeley National Laboratory, Berkeley, USA

E-mail address: kapersson@lbl.gov

[c] Professorship for Renewable Energy Carriers, ETH Zurich, Zurich, Switzerland

E-mail address: bulfinb@ethz.ch

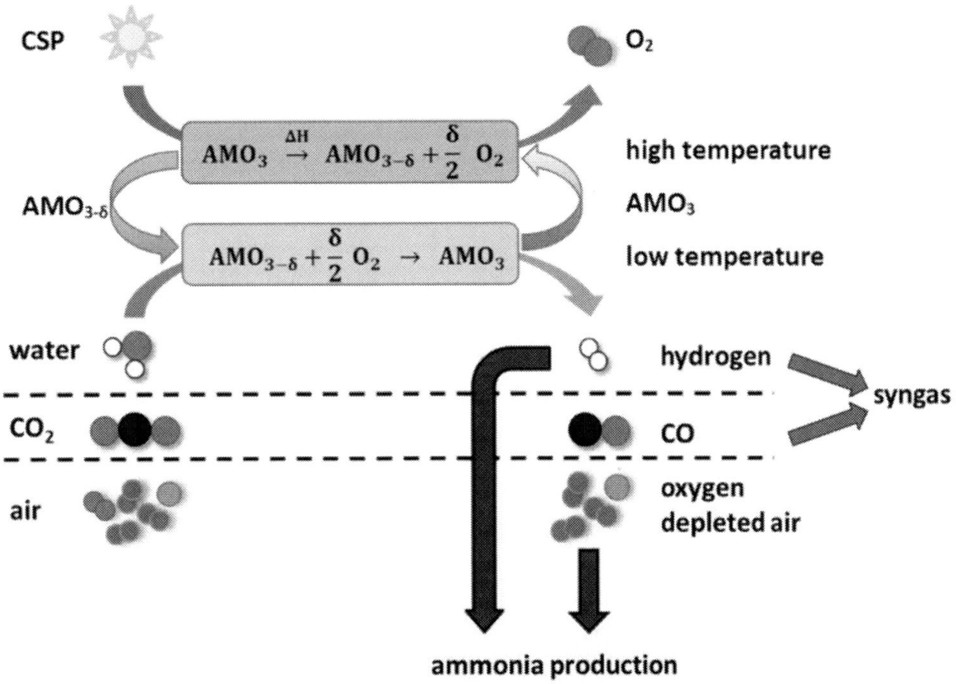

Fig. 1: Thermochemical cycles for the production of solar fuels or for air separation and ammonia production. Adapted from [1].

Converting heat from renewable sources into other forms of energy is considered an essential factor in the reduction of greenhouse gas emissions. For instance, high temperatures can be reached using concentrated solar power (CSP), and the thus-captured energy can be converted into so-called solar fuels via thermochemical processes. These consist of the partial reduction of a redox material, usually a metal oxide, at high temperatures following the exothermic re-oxidation of this material at a lower temperature level using steam or CO_2, which are thus converted into hydrogen or carbon monoxide, respectively. These two gases can be combined to generate syngas for the production of hydrocarbons (see Fig. 1). Through the same process, a stream of mostly inert gas can be produced by re-oxidation with air, allowing air separation using renewable energy sources. Hydrogen production and air separation can also provide the feedstock for ammonia production through the Haber-Bosch process, as the achieved oxygen partial pressures can be kept low enough to avoid catalyst poisoning. [2] Ammonia produced through this method can be used for fertilizer production, or as a fuel for energy storage.

Achieving efficient air separation and fuels production through solar-thermochemical processes is challenging but possible. Finding suitable redox materials depending on the respective process conditions through evaluation of the materials thermodynamics is a key point in reaching high process efficiencies. [1, 3-5] Within a materials screening for these applications, we prepared perovskite solid solutions with the general composition $A_xB_{1-x}M_yN_{1-y}O_{3-\delta}$ with A, B = Ca, Sr and M, N = Ti, Mn, Fe, Co, Cu using a modified Pechini method. [5] Their redox enthalpies and entropies as a function of the non-stoichiometry δ can be tuned by adjusting their composition. We obtained experimental data gathered via equilibrium oxygen uptake and release measurements using thermogravimetric analysis, and theoretical data gathered via density functional theory (DFT) calculations. The experimental data, i.e., redox enthalpies and entropies, are fit using a novel empirical model, in order to generate interactive isotherms, isobars, as well as graphs at constant non-stoichiometry which are referred to as isoredox plots (see Fig. 2). In a joint effort between the *German Aerospace Center* and the Lawrence *Berkeley National Laboratory*, the data is used to create a search engine for redox materials data based upon the infrastructure of *The Materials Project*. [6] The data is included into *MPContribs* [7], which is the framework for external contributors to publicly share their data on the Materials Project website. Many of the functions included in this contribution are based on the databases included in *The Materials Project*.

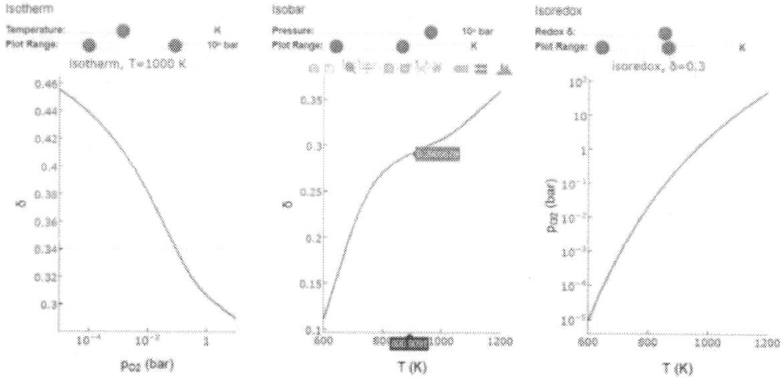

Fig. 2: Interactive isotherms, isobars, and "isoredox" plots as they are integrated into *MPContribs* (here for $SrFeO_{3-\delta}$).

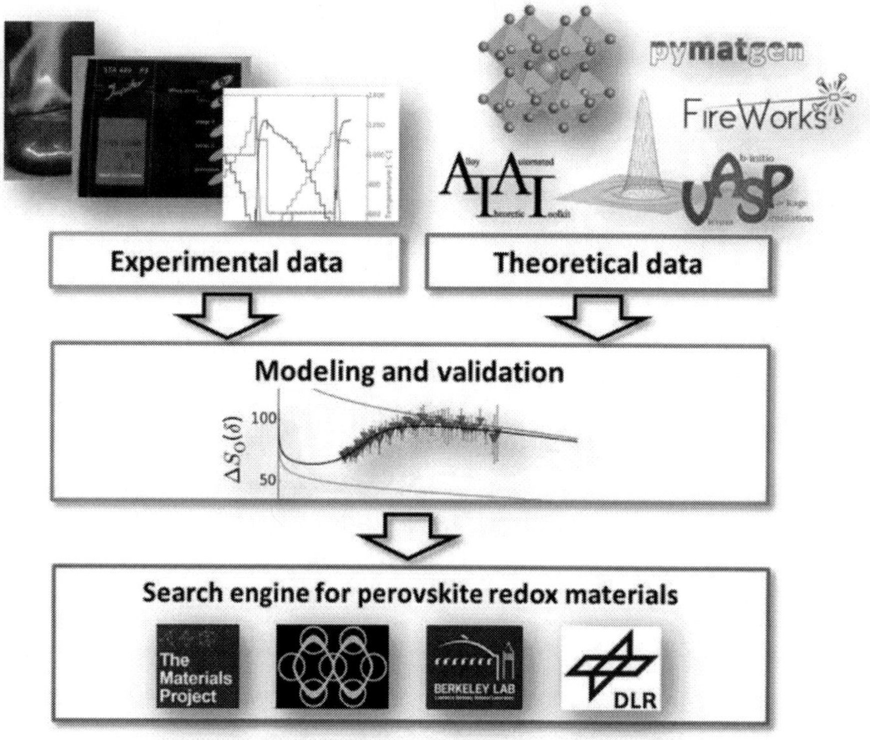

Fig. 3: Combination of experimental and theoretical data within The Materials Project.

Moreover, theoretical data is collected for a large set of perovskite redox materials, including solid solutions in a large compositional range. This data is generated by preevaluating possible stable candidate materials using the Goldschmidt tolerance factor and subjecting this data to Density Functional Theory (DFT) based calculations. The resulting structural data contains the energies of all atoms in the structure and allows to calculate redox enthalpies, and, via the elastic tensors of the materials and their composition, redox entropies. Thus, a full model of thermodynamic properties is generated. Both experimental and theoretical datasets are used to create a redox materials database. By defining specific target process conditions, such as the reduction and oxidation temperatures and oxygen partial pressures, it is possible to find the most efficient redox material for each specific application using the novel perovskite search engine.

These factors, in summary, accelerate the finding of new materials by replacing large sets of experiments by a computer-based pre-selection step. By finding many new redox materials and selecting the best candidates for further studies, we allow a major leap towards more efficient renewable energy conversion and storage, including ammonia production and solar fuels generation.

References

[1] J. Vieten, B. Bulfin, F. Call, M. Lange, M. Schmuecker, A. Francke, M. Roeb, C. Sattler, Journal of Materials Chemistry A, 4, 13652-13659 (2016)

[2] B. Bulfin, J. Lapp, D. Guban, J. Vieten, S. Richter, S. Brendelberger, M. Roeb, C. Sattler, submitted, (2018)

[3] J. Vieten, B. Bulfin, M. Senholdt, M. Roeb, C. Sattler, M. Schmuecker, Solid State Ionics 308, 149-155 (2017)

[4] B. Bulfin, J. Vieten, C. Agrafiotis, M. Roeb, C. Sattler, Journal of Materials Chemistry A, 5, 18951-18966 (2017)

[5] J. Vieten, B. Bulfin, M. Roeb, C. Sattler, Solid State Ionics 315, 92-97 (2018)

[6] A. Jain, S.P. Ong, G. Hautier, W. Chen, W.D. Richards, S. Dacek, S. Cholia, D. Gunter, D. Skinner, G. Ceder, K.A. Persson, APL Materials, 011002 (2013)

[7] https://materialsproject.org/mpcontribs, *RedoxThermoCSP* contribution under construction.

527376 Microwave Catalysis for Ammonia Synthesis Under Mild Reaction Conditions

Abstract:

Microwave Catalysis for Ammonia Synthesis under Mild Reaction Conditions

Jianli Hu, Hanjing Tian, Yan Luo, Xinwei Bai, West Virginia University

Dushyant Shekhawat, Christina Wildfire, Victor Abdel-Sayed, Michael Spencer, National Energy Technology Laboratory

Robert Dagle, Stephen Davidson, Pacific Northwest National Laboratory

Al Stiegman, Florida State University

A scalable, cost-effective catalytic process of ammonia synthesis is developed by using microwave excitation under mild reaction conditions. In this research project funded by DOE ARPA-E, our interdisciplinary team of WVU, NETL, PNNL, FSU and two industrial partners have demonstrated that ammonia synthesis can be carried out at 200-300°C and ambient pressure. This transformational process integrates system elements of electromagnetic sensitive catalysts and microwave reactor design. Taking advantages of ¡°state-of-the art¡± non-equilibrium microwave plasma technology, catalytic ammonia synthesis undergoes a new reaction pathway where the barrier for the initial dissociation of the dinitrogen is decoupled from the bonding energy of the intermediates.

In this project, catalytic ammonia synthesis was conducted under two microwave irradiation scenario: microwave and microwave plasma. Other than selective activation of dinitrogen to metastable radicals, the most obvious advantage that microwave irradiation affords in driving a heterogeneously catalyzed reaction is the ability to locally heat the catalytic sites. Many industrial processes utilizing heterogeneous catalysts are high-temperature processes wherein both components of the reaction (i.e., catalyst and medium) are heated to the temperature required for the reaction to occur.

Microwave irradiation is a facile and efficient means of generating plasmas and has been used for that purpose in a number of applications. The transformational chemistry is driven by the specific characteristics of plasma chemistry that leads to an energy efficient paradigm for the synthesis of ammonia. It is equally important to note that microwave irradiation, regardless of its accompanying plasma generation, has shown a profound impact on catalyzed and uncatalyzed gas-solid reactions. In particular, it has been demonstrated that microwave-specific effects can manifest themselves through the enhancement of reaction rates, changes in the position of equilibria and the distribution of products.

Specifically discussed in this presentation are the effects of electromagnetic properties of catalysts, microwave frequency and microwave energy absorbed on the conversion of nitrogen and ammonia yield. Energy efficiency and Technoeconomic analysis (TEA) relevant to microwave catalytic synthesis of ammonia will also be discussed to shed lights on commercial viability.

Figure 1. Illustration of microwave catalytic synthesis of ammonia

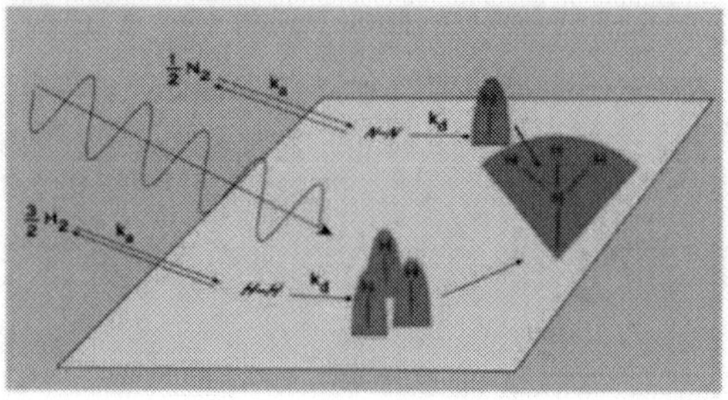

539911 Basic Co-Firing Characteristics of Ammonia with Pulverized Coal in a Single Burner Test Furnace

Abstract:

Ammonia is expected as a potential fuel to substitute fossil fuels, because it does not discharge carbon dioxide and is easily handled by liquefaction. There are several ways for the direct use of ammonia as a fuel; for example, use in fuel cells and combustion devices. One of the possible àpplication is the combustion use in thermal power plants. In particular, co-firing of ammonia in coal-fired power plants seems to have a relatively great advantage on the suppression of greenhouse gases, because coal is one of the main emission source of carbon dioxide. On the other hand, it is concerned that concentration of nitrogen oxides (NO_X), which is one of the typical atmospheric pollutant, in the flue gas would considerably increase due to the oxidation of ammonia. To utilize ammonia as a co-firing fuel in existing pulverized coal-fired power plant, without causing additional costs for the modification of the denitration equipment, it is important to develop a combustion technology that can suppress the NO_x concentration in the flue gas. Co-firing characteristics of pulverized coal and ammonia, however, had not been evaluated except in the case of very low co-firing rate for the purpose of denitration in the pulverized coal flame. In this study, basic co-firing characteristics of pulverized coal and ammonia were investigated using a bench-scale single burner test furnace.

A schematic diagram of the single burner test furnace is shown in Fig. 1. The rated thermal input of the furnace is 760 kW, which corresponds to approximately 100 kg/h of coal for common bituminous coals. The total air ratio (the inverse of the equivalence ratio) is 1.24, in which the oxygen concentration at the furnace exit is 4%. In addition, two-stage combustion is employed to reduce NO_x emission by creating a reducing environment, and the staged air was fed from the ports located at a distance of 3.0 m from the burner. The rate of staged air is 30% of the total air flow. Ammonia was fed into the furnace in two different ways; one is the injection into the center of the pulverized coal burner, and the other is separate injection through the side port of the furnace wall.

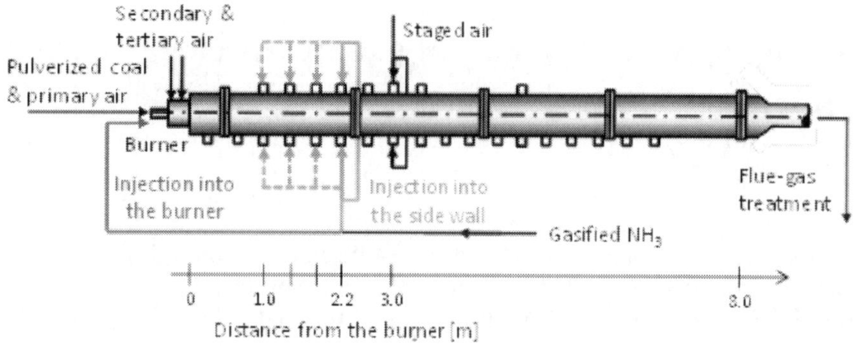

Fig. 1. Schematic of the combustion test furnace.

First, we examined the effects of ammonia co-firing rate and injection method on NO_x concentration in the flue gas. Fig. 2 shows the effect of ammonia co-firing rate on NO_x concentration in the flue gas in the case of ammonia injection into the burner. There was no significant change in NO_x concentration in the case of 5 and 10% co-firing rate compared to the case of single coal combustion. However, as increasing co-firing rate to 15 and 20%, NO_x concentration also increased, and the concentration at co-firing rate of 20% was approximately 20% higher than that at single coal combustion. This indicates that fuel NO_x was produced in the flame by the oxidation of ammonia, or combustion of pulverized coal was aggravated, and then it resulted in the increase of NO_x emission. On the other hand, it was found that NO_x concentration in the flue gas at 20% ammonia co-firing depended on the position of ammonia injection into the furnace. Fig. 3 shows the changes of NO_x concentration against different locations of ammonia injection, where ammonia co-firing rate is constant at 20%. When ammonia was injected through the side port which is located at 1.0 m apart from the burner, NO_x concentration decreased compared to the case of ammonia injection into the burner, and was almost equal to that for single coal combustion. When ammonia was injected through the port located at 1.4, 1.8 or 2.2 m apart from the burner, however, NO_x concentration became higher as the injection position was shifted to downstream, and the NO_x concentration exceeded that for single coal combustion. The reason that the NO_x concentration did not increase when ammonia was injected through the port at 1.0 m from the burner is expected the denitration effect of ammonia on NO_x generated in pulverized coal flame. Therefore, by injecting ammonia into the region of low-O_2 and high-NO_x concentration, it is expected that not only the generation of fuel-NO_x is effectively suppressed, but also ammonia can work as a denitration agent against the existing NO_x in the flame.

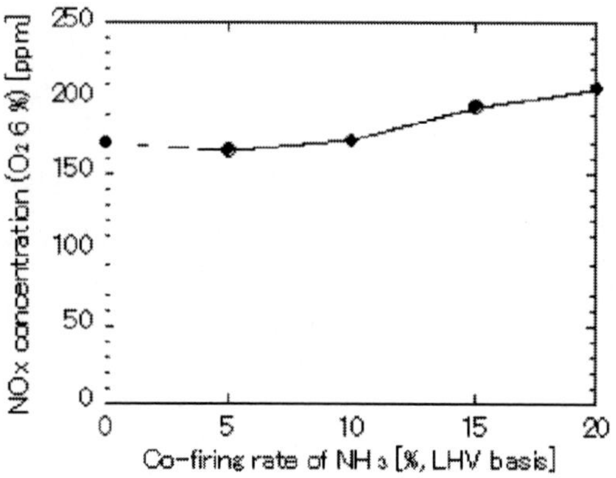

Fig. 2. Effect of ammonia co-firing rate on NO_x concentration in the flue gas when ammonia was injected into the pulverized coal burner.

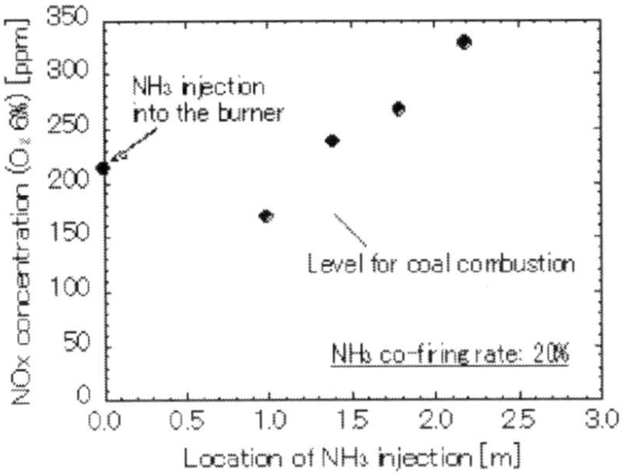

Fig. 3. Effect of the location of ammonia injection (distance from the burner) on NO_x concentration in the flue gas.

The influence of ammonia co-firing on unburned carbon concentration in the fly ash and concentration of ammonia and nitrous oxide (N_2O) in the flue gas, where N_2O is one of the strongest greenhouse gases, was also evaluated. When ammonia was injected into the center of the burner with the co-firing rate of 20%, unburned carbon concentration in the fly ash increased by approximately 20%. In this case, flame temperature near the burner was lower compared to the case of single coal combustion, and it presumably affected the increase of unburned carbon concentration in the fly ash. Concentration of unburned ammonia and N_2O in the flue gas was also a little higher in this case, but these increases were less than a few ppm and 1 ppm, respectively. Since ammonia is also used in the denitration equipment with selective catalytic reduction process, increase of unburned ammonia by a few ppm would not cause significant problem on the actual coal-fired power plants. Also, increase of N_2O by a few ppm is almost negligibly low to work as a greenhouse gas, because emission of carbon dioxide is decreased by almost 20% in the 20% ammonia co-firing condition. In the case ammonia was injected through the side port at 0.6 or 1.0 m from the burner, unburned carbon in the fly ash, unburned ammonia and N_2O in the flue gas did not increase compared to the case of single coal combustion; therefore it is considered that the significant problem would not occur, either in this case. As a result, we could have demonstrated the possibility of ammonia co-firing up to 20% in the coal-fired thermal power plants.

This research was supported by a program of Energy Carrier in Strategic Innovation Promotion Program (SIP), led by Japan Science and Technology Agency (JST). We appreciate the support from all concerned.

540285 Development of Low-NO_x Combustor of Micro Gas Turbine Firing Ammonia Gas

Abstract:

Abstract

A massive influx of renewable energy is required in order to mitigate global warming. Although hydrogen is a renewable media, its storage and transportation in large quantity is difficult. Ammonia, however, is a hydrogen energy carrier, and its storage and transportation technology is already established. Although ammonia fuel combustion was studied in the 1960s in the USA, the development of an ammonia fuel gas turbine had been abandoned because combustion efficiency was unacceptably low [1]. Recent demand for hydrogen energy carrier revives the usage of ammonia fuel. The National Institute of Advanced Industrial Science and Technology (AIST) in Japan, in collaboration with Tohoku University successfully realized ammonia-kerosene gas turbine power generation in 2014, and ammonia fuel gas turbine power generation in 2015 by using 50-kWe class gas turbine [2, 3].

The drawback of the facility is that it requires a large-size selective catalytic reduction (SCR) to decrease the high concentrations of NOx. In order to promote the widespread of ammonia combustion gas turbine system, it is necessary to downsize SCR NOx reduction. In other words, it is important to reduce NOx emission from ammonia gas turbine combustor. AIST has begun developing low-NOx combustors by using a combustor test rig. Simultaneously, fundamental research was carried out at Tohoku Univ. It was found at Tohoku Univ. that rich-lean two-stage combustion method and a control of equivalence ratio of the primary combustion zone to around the value of 1.1 to 1.2 significantly decreases NO emissions in gas-turbine swirl combustor [4, 5].

Based on the knowledge of rich-lean combustion method, the flow paths of the combustor in the combustor rig tests were systematically varied to make a fuel rich region in the primary combustion zone. First, NOx emissions of standard combustor was measured in the range of fuel input energy of 140 to 330 kJ/s where NH3 fuel input energy of 236 kJ/s was over the limit that can be supplied by an NH3 fuel supply facility. Although, input energy over 236 kJ/s was supplied by CH4 fuel, it was shown in the previous study [3] that the addition of NH3 in the CH4+NH3 mixture increased the NO emissions for the most part. NO emission of standard combustor was flat 1000 (16% O2) ppm from 140 to 330 kJ/s of CH4+NH3 fuel input energy. Next, NOx emissions of the modified combustors were measured. The modifications applied to the combustor included filling the primary dilution holes by using stainless bolts, opening the secondary dilution holes by drilling, and halving the air swirler area by using stainless hook or stainless wire reel. These modifications were intended to increase the equivalence ratio of primary combustion zone. Valley of NO emissions of the modified combustor was observed in the range of fuel input energy of 140 to 236 kJ/s, which implied the ability of low-NOx gas turbine combustion firing ammonia gas.

Acknowledgements

This work was supported by the Council for Science, Technology and Innovation (CSTI), the Cross-ministerial Strategic Innovation Promotion Program (SIP), "Energy Carriers" (Funding agency: Japan Science and Technology Agency (JST)).

References

[1] Pratt, D.T., Performance of Ammonia-fired Gas-turbine Combustors, Technical Report No.9, DA-04-200-AMC-791(x), Berkley University of California (1967), available at < http://www.dtic.mil/dtic/tr/fulltext/u2/657585.pdf>.

[2] Iki, N., Kurata, O., Matsunuma, T., Inoue, T., Suzuki, M., Tsujimura, T., Furutani, H., Kobayashi, H., Hayakawa, A., Arakawa, Y., Ichikawa, A., Micro Gas Turbine Firing Ammonia, The 12th Annual NH3 Fuel Conference, Chicago, September 20-23, (2015), available at < https://nh3fuelassociation.org/conference2015/>.

[3] Kurata, O., Iki, N., Matsunuma, T., Inoue, T., Tsujimura, T., Furutani, H., Kobayashi, H., Hayakawa, A., Performances and Emission Characteristics of NH3-air and NH3-CH4-air Combustion Gas-turbine Power Generations, *Proc. Combust. Inst.*, **36** (2017) 3351-3359.

[4] A. Hayakawa, Y. Arakawa, R. Mimoto, K. D. Kunkuma A. Somarathne, T. Kudo, H. Kobayashi, *Int. J. Hydrogen Energy*, **42** (2017) 14010–14018.

[5] K. D. Kunkuma A. Somarathne, S. Hatakeyama, A. Hayakawa, H. Kobayashi, *Int. J. Hydrogen Energy*, **42** (2017) 27388–27399.

TWO STAGE AMMONIA COMBUSTION IN A GAS TURBINE LIKE COMBUSTOR FOR SIMULTANEOUS NO AND UNBURNT AMMONIA REDUCTIONS

Dr. Akihiro Hayakawa, Institute of Fluid Science, Tohoku University, JAPAN
Dr. K.D. Kunkuma A. Somarathne, Institute of Fluid Science, Tohoku University, JAPAN
Mr. Masaaki Tsukamoto, Institute of Fluid Science, Tohoku University, JAPAN
Mr. Taku Kudo, Institute of Fluid Science, Tohoku University, JAPAN
Prof. Hideaki Kobayashi, Institute of Fluid Science, Tohoku University, JAPAN

Introduction

Recently the use of ammonia is anticipated not only as hydrogen energy carrier but also as carbon-free fuel. In order to employ ammonia as fuel for an internal combustion engine, such as reciprocating engine or gas turbine combustor, fundamental flame characteristics need to be clarified. Hayakawa et al. (1) clarified the laminar burning velocities and the Markstein lengths up to elevated pressure conditions. The unstretched laminar burning velocity of ammonia/air mixture at atmospheric pressure is about 7 cm/s which is about 1/5 of that of methane/air flame, and the value decreases with an increase in the mixture pressure. The addition of hydrocarbon to ammonia is one of the options in order to enhance the weak flame characteristics of ammonia flame. Okafor et al. (2) studied the laminar burning velocities of methane/ammonia binary fuel for various equivalence ratios and methane/ammonia mixing rate conditions, and proposed an improved detailed reaction kinetics that is validated by the experimental values of laminar burning velocity.

Since the swirl combustor is widely employed in a gas turbine combustor to improve flame stability characteristics, fundamental study of ammonia flames in swirling flow is important to realize ammonia-fueled gas turbine combustor. Hayakawa et al. (3) experimentally showed that the ammonia flame can be stabilized for various equivalence ratio range and mixture inlet velocity. Valera-Medina et al. showed the product gas characteristics of swirl stabilized amnonia/methane flames (4).

Kurata et al. (5) successfully demonstrated the power generation of 41.8 kW at the Fukushima Renewable Energy Institute, AIST (FREA) by ammonia and methane-ammonia firing. In Kurata's demonstration, the combustor was a modified combustor of the burner for kerosene combustion, and the design was not optimized for the ammonia combustion. Therefore, more than 500 ppm of NO was generated downstream of the combustor. Although NO and unburnt ammonia could be reduced by adopting a selective catalytic reduction (SCR) system, further reduction of NO and ammonia needs to be reduced in order to reduce the size of SCR system is required.

Recently, Somarathne et al. (6) numerically proposed a combustion concept of "the rich-lean two stage combustion" to reduce unburnt ammonia and NOx simultaneously. The three-dimensional large eddy simulation with detailed reaction kinetics showed that the amount of NO and unburnt ammonia can be reduced simultaneously by controlling the equivalence ratio of primary zone. The rich-lean two-stage combustion concept is also useful for an actual gas turbine combustor (7, 8). In the two-stage combustion concept, equivalence ratios in the primary zone and the total (overall) equivalence ratio changes the product gas characteristics. Therefore, in this study, product gas characteristics of gas-turbine like combustor was experimentally investigated. The rich-lean two-stage combustion concept was employed in the combustor. Then, effects of primary zone equivalence ratios, overall equivalence ratios on product gas characteristics are experimentally investigated.

Experimental Setup

Figure 1 shows the swirl burner used in this study. An axial type swirl combustor was employed in this study as shown in Figs. 1a and 1b. The swirl burner was also employed in our previous study (3). The dimensions of the swirler and inner diameter of the liner were bigger than the burner used and the previous burner was not water cooled in previous study (3). Therefore, the burner might be damaged by the heat release, especially at high pressure conditions. In the present burner shown in Fig. 1a, the scale of the swirler was smaller than the previous one. Therefore, the heat release inside the burner can be reduced. In addition, the water burner was water cooled. The water temperature was set to 60 °C. The consistency of temperatures of the bottom plate surface and the bluff body is considered to be better than the previous swirl burner.

A swirler is mounted at the bottom of the burner. The outer and inner diameters of the swirler were 24 mm and 14 mm, respectively, and the angle of swirler vane was 40°. Therefore, the swirl number determined from the dimensions of the swirler was 0.68. A quartz glass liner was placed on the bottom plate as a burner liner. Inner diameter and the height of the liner were 72 mm and 150 mm, respectively, and a top plate with the exhaust outlet hole whose diameter was 36 mm was placed. These dimensions were the same with the calculation domain of three-dimensional large eddy simulation by Somarathne et al. (6).

Ammonia was used as fuel and dry air was employed as oxidizer. Ammonia/air pre-mixture was supplied from the bottom of the burner through the swirler. The equivalence ratio, ϕ, of the mixture can be determined. As mentioned later, the equivalence ratio in the primary zone, ϕ_p, also calculated by the equation below,

$$\phi, \phi_p = \frac{Q_{p,f}/Q_{p,a}}{\left(Q_{p,f}/Q_{p,a}\right)_{st}} = 3.57 \cdot \frac{Q_{p,f}}{Q_{p,a}}, \tag{1}$$

here, $Q_{p,f}$, $Q_{p,a}$ represent the flow rates of fuel and air through the swirler, and the subscript "st" represents the condition at the stoichiometry. The mixture inlet velocity was set for 4.0 m/s for all experiments.

In this study, the high pressure combustion experiments were also performed using the high pressure combustion test facility at the Institute of Fluid Science, Tohoku University. The details of the high pressure test facility have been described elsewhere (9).

A secondary air injection was also carried out in this study in order to reduce NO and unburnt ammonia simultaneously. Figs. 1c and 1d shows the swirl burner with secondary air injection. The air injector has 8 injection ports, and the size of air injection port was 10 mm x 10 mm. The height of the center of secondary air injector was 100 mm above from the burner bottom. The geometry of the secondary injector also the same with the three dimensional numerical simulation by Somarathne et al. (6).

When the secondary air injection was employed, the overall equivalence ratio, ϕ_o, also can be determined. Considering the Eq. (1), the overall equivalence ratio was evaluated from Eq. (2),

$$\phi_o = 3.57 \cdot \frac{Q_{p,f}}{Q_{p,a}+Q_{s,a}}, \tag{2}$$

here, $Q_{s,a}$ was the total flow rate of air injected through the secondary air injector. At the outlet of the secondary air injector, a glass wool sheet was placed for uniform air injection.

For burnt gas analysis, a Fourier Transform Infrared (FTIR) gas analyzer combined with O_2 and H_2 analyzer (Best Sokki, BOB-2000FT) was employed. The accuracy of the FTIR analyzer was 1% to

2% of full scale depending on the species. Figure 1e shows the schematic of the product gas sampling system. Product gas at the burner exit is not uniform. In order to measure a spatially averaged amount of product throughout the burner exit, a mixing plate was mounted at the burner exit as shown in Fig. 1f. Figure 1g shows the detail of the mixing plate. The sampling point was at 5 mm or 100 mm downstream of the height of the mixing plate outlet. In order to reduce the influence of progress of reaction inside the sampling probe, a triple tube prove with air cooling was employed. The temperature inside the probe maintained above 100 °C in order to prevent from the condensation of water.

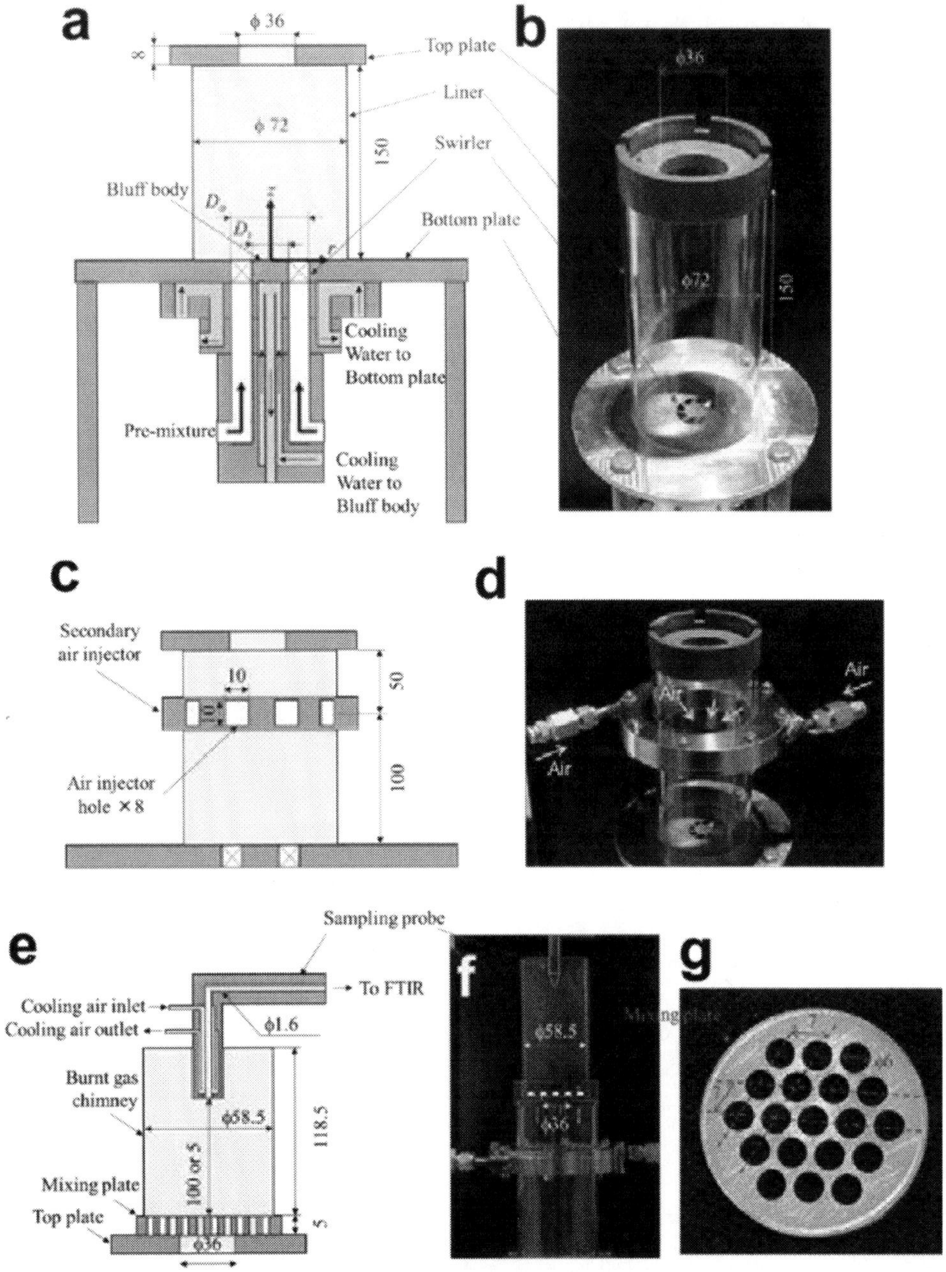

Fig. 1 Experimental setup: (a) schematic of water-cooled burner; (b) photo of water-cooled burner; (c) schematic of secondary air injector; (d) photo of secondary air injector; (e) schematic of product gas sampling system; (f) photo of product gas sampling system and (g) photo of the mixing plate.

Results and Discussions

Product Gas Characteristics without Two Stage Combustion

Figure 2 shows NO, unburnt ammonia and hydrogen concentrations in the product gas from the swirl combustor for various mixture pressure. In this figure, the measurement point was the center of the burner exit and the height was 5 mm above the top plate, and the mixing plate was not installed. As mentioned above, the composition of the product gas at the burner exit was not uniform because of the influence of non-adiabatic quartz liner wall. At this sampling point, the composition of the product gas is considered to be close to the flame with no chemical quenching. As shown in Fig. 2a, NO concentration increased with a decrease in the equivalence ratio. In lean condition, NO concentration decreased with an increase in pressure. The trend of the decrease in NO with an increase in the equivalence ratio and pressure was explained by Hayakawa et al. (10). The decrease in NO with an increase in equivalence ratio is due to the decrease in OH concentration that plays an important role in NO production through fuel NO and the reduction of NO by unburnt ammonia. The OH concentration also decreases with an increase in pressure because OH consumption rate increases through the third-body reaction, OH + H + M <=> H_2O + M, and therefore NO production rate at high pressure is reduced. The unburnt ammonia and hydrogen concentration rapidly increases with an increase in equivalence ratio at rich conditions.

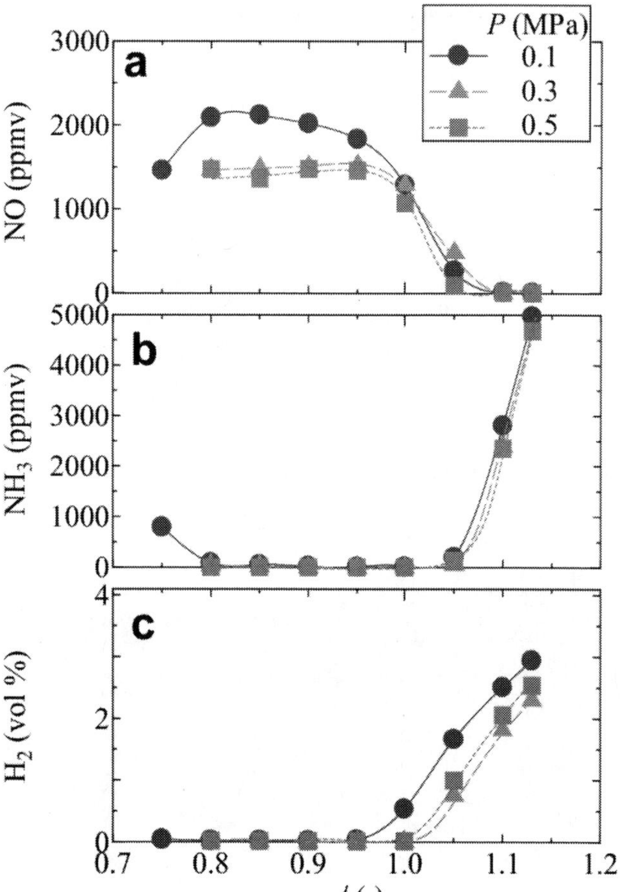

Fig. 2 Product gas characteristics with no mixing plate

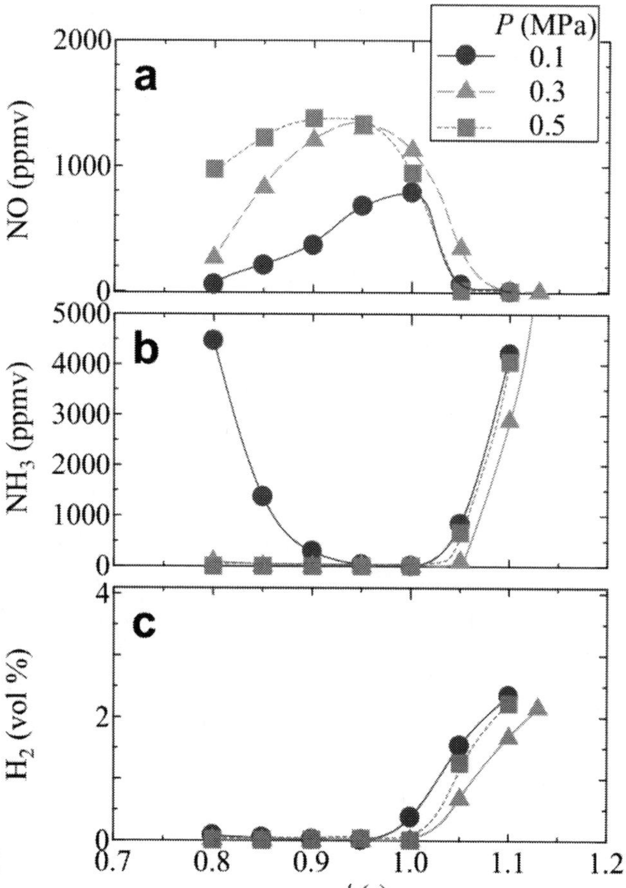

Fig. 3 Product gas characteristics with mixing plate

Figure 3 shows the product gas characteristics with the mixing plate and the sampling point was 100 mm downstream of the mixing plate. The mixture compositions described here was considered to be a space-averaged product gas throughout the burner outlet although the residence time of the product gas was different. In this case, the NO concentration at 0.1 MPa were lower than that at other pressure conditions. It was considered that the reaction was quenched due to the heat loss to the liner wall, and the unburnt ammonia was mixed with other product gas that includes NO. Small amount of NO is due to the reduction by unburnt ammonia. As equivalence ratio decreased, the characteristic time of chemical reaction becomes long and the amount of unburnt ammonia round the liner was considered to increase because of slower laminar burning velocity. Therefore, the amount of unburnt ammonia rapidly increased for small equivalence ratio case. Even in this case, the amount of NO and unburnt ammonia reaches minimum around $\phi = 1.05$.

Simultaneous Reduction of NO and Ammonia using Lean-Rich Two Stage Combustion

Figure 4 shows the product gas characteristics 100 mm downstream of the mixing plate with secondary air injection at $P = 0.1$ MPa for various primary zone equivalence ratio, ϕ_p. Here, overall reaction was set to $\phi_o = 0.5$. In this case, NO concentration slightly decreased from the case of no-secondary air injection shown in Fig. 3 but unburnt ammonia concentration increased, especially at stoichiometric condition. In the case of atmospheric pressure, the flame length was relatively long and the reaction could not complete by the height of secondary air injection. Therefore, the ammonia reaction was quenched by the secondary air. Therefore, enough length for complete reaction is required in order to reduce the amount of unburnt ammonia by secondary air injection.

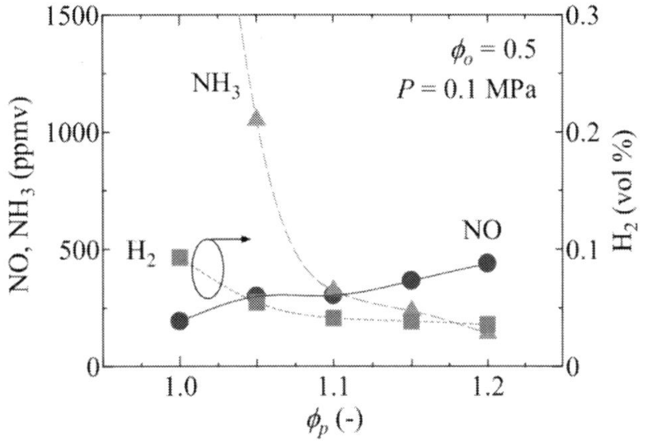

Fig. 4 Effects of primary zone equivalence ratio, ϕ_p, on the product gas characteristics of NO, unburnt NH_3 and H_2 at mixture pressure of 0.1 MPa

Since the decrease in the chemical reaction time as pressure increases, the flame height decreases with an increase in pressure. Figure 5 shows the product gas characteristics at 0.5 MPa with secondary air injection for various primary zone equivalence ratio. In this case, the amount of unburnt ammonia and the hydrogen concentration was almost zero, i.e., unburnt ammonia and hydrogen burned completely by the secondary air. As shown in Fig. 5, there was an optimum primary zone equivalence ratio to reach the minimum amount of NO. The same trend was also observed in the numerical simulation by Somarathne et al. (6) and experimental study by Okafor et al. (7). In this experiment, the optimum value of primary zone equivalence ratio was 1.05 and is slightly rich condition. As decrease in primary zone equivalence ratio, the NO concentration produced in primary zone increases and the amount of unburnt ammonia decreases as shown in Figs. 2 and 3. Therefore, no reaction was occurred by secondary air injection. On the other hand, since the amount of unburnt ammonia and hydrogen increased drastically as primary zone equivalence ratio increases as shown in Figs. 2 and 3, OH radical was considered to be generated when hydrogen burns. The increase in OH radical enhanced NO production through HNO production.

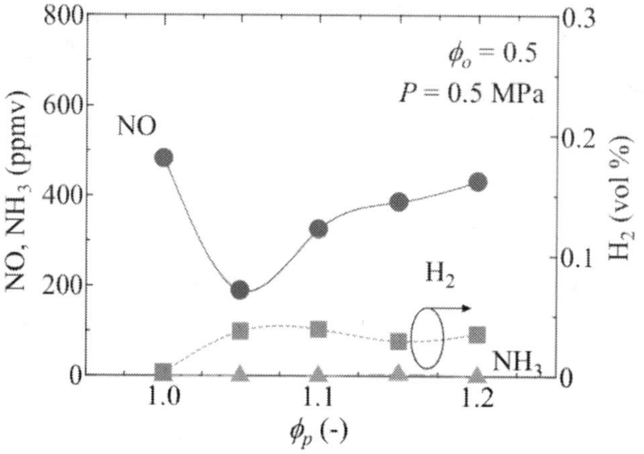

Fig. 5 Effects of primary zone equivalence ratio, ϕ_p, on the product gas characteristics of NO, unburnt NH_3 and H_2 at mixture pressure of 0.5 MPa

Figure 6 shows the effect of overall equivalence ratio, ϕ_o, on the amount of product gas. The Japanese regulatory limit of NOx is determined based on the case at 16% O_2. Therefore, the amount of NO, unburnt ammonia and H_2 was converted to the values at 16% O_2. The broken line of 70 ppmv of NO corresponds the Japanese regulatory limit of NOx from gas turbine. Although the product gas characteristics changes with the overall equivalence ratio but not so sensitive as primary zone equivalence ratio. Therefore, the control of primary zone equivalence ratio is m important than that of overall equivalence ratio. In the cases of ϕ_o = 0.35 and 0.4, the amount of NO was much lower than 70 ppm. Therefore, secondary air injection has potential of simultaneous reduction of ammonia from gas turbine combustor.

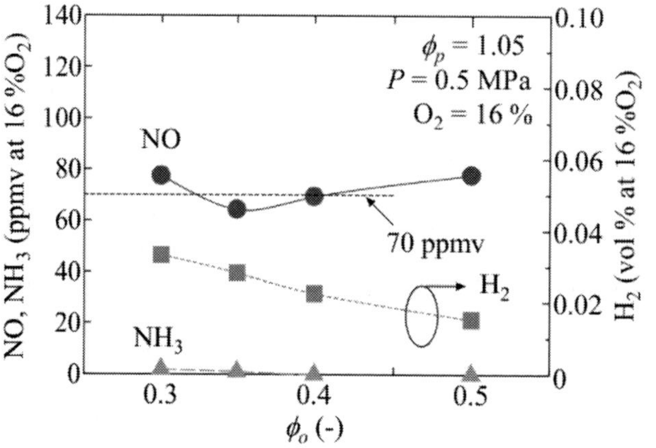

Fig. 6 Effects of overall equivalence ratio, ϕ_o, on the product gas characteristics of NO, unburnt NH_3 and H_2 at 16% O_2 conditions under 0.5 MPa

Conclusions

In order to achieve simultaneous reductions of NO and unburnt ammonia, the product gas characteristics of swirl burner with secondary air injection was experimentally investigated. As results, the following results were obtained:

1. The position of the secondary air injection needs to place downstream of the main reaction zone of primary zone. The amount of unburnt ammonia increases when the secondary air is injected to the reacting area due to the quenching of reaction.
2. There is an optimum primary zone equivalence ratio for simultaneous NO and unburnt ammonia reduction, and primary zone equivalence ratio is sensitive to the amount of NO. In this experiment, the optimum primary zone equivalence ratio was 1.05. Therefore, control of primary zone equivalence ratio is quite important in an actual ammonia-fueled gas turbine.
3. The NO can be reduced below the Japanese emission regulation with no unburnt ammonia by applying secondary air injection. The overall equivalence ratio also affects to the NO characteristics but less sensitive compared to the primary zone equivalence ratio.

Acknowledgment

This study was supported by the Council for Science, Technology and Innovation (CSTI), the Cross-ministerial Strategic Innovation Promotion Program (SIP), "Energy Carriers" (Funding Agency: the Japan Science and Technology Agency (JST)).

References

1. Hayakawa, A., Goto, T., Mimoto, R., Arakawa, Y., Kudo, T. and Kobayashi, H. (2015), “Laminar Burning Velocity and Markstein Length of Ammonia/air Premixed Flames at Various Pressures,” *Fuel*, Vol. 159, pp. 98-106.
2. Okafor, E.C., Naito, Y., Colson, S., Ichikawa, A., Kudo, T., Hayakawa, A. and Kobayashi, H. (2018), “Experimental and Numerical Study of the Laminar Burning Velocity of CH_4-NH_3-air Premixed Flames,” *Combustion and Flame*, Vol. 187, pp. 185-198.
3. Hayakawa, A., Arakawa, Y., Mimoto, R., Somarathne, K.D.K.A., Kudo, T. and Kobayashi, H. (2017), “Experimental Investigation of Stabilization and Emission Characteristics of Ammonia/air Premixed Flames in a Swirl Combustor,” *International Journal of Hydrogen Energy*, Vol. 42, pp. 14010-14018.
4. Valera-Medina, A., Marsh, R., Runyon, J., Pugh, D., Beasley, P., Hughes, T. and Bowen, P. (2017), “Ammonia-methane Combustion in Tangential Swirl Burners for Gas Turbine Power Generation,” *Applied Energy*, Vol. 185, pp. 1362-1371.
5. Kurata, O., Iki, N., Matsunuma, T., Inoue, T., Tsujimura, T., Furutani, H., Kobayashi, H. and Hayakawa, A. (2017), “Performances and Emission Characteristics of NH3-air and NH3-CH4-air Combustion Gas-turbine Power Generations,” *Proceedings of the Combustion Institute*, Vol. 36, pp. 3351-3359.
6. Somarathne, K.D.K.A., Hatakeyama, S., Hayakawa, A. and Kobayashi, H. (2017), “Numerical Study of a Low Emission Gas Turbine Like Combustor for Turbulent Ammonia/air Premixed Swirl Flames with a Secondary air Injection at High Pressure,” *International Journal of Hydrogen Energy*, Vol. 42, pp. 27388-27399.

7. Okafor, E.C., Somarathne, K.D.K.A., Hayakawa, A., Kudo, T., Kurata, O., Iki, N., Kobayashi, H. (2019), "Towards the Development of an Efficient Low-NOx Ammonia Combustor for a Micro Gas Turbine," *Proceedings of the Combustion Institute*, Vol. 37, in press.
8. Kurata, O., Iki, N., Inoue, T., Matsunuma, T., Tsujimura, T., Furutani, H., Kawano, M., Arai, K., Okafor, E.C., Hayakawa, A. and Kobayashi, H. (2019), "Development of a Wide Range-operable, Rich-lean Low-NOx Combustor for NH_3 Fuel Gas-Turbine Power Generation," *Proceedings of the Combustion Institute*, Vol. 37, in press.
9. Kobayashi, H., Tamura, T., Maruta, K., Niioka, T. and Williams, F.A. (1996), "Burning Velocity of Turbulent Premixed Flames in a High-pressure Environment," *Twenty-Sixth ymposium (International) on Combustion*, Vol. 26, pp. 389-396.
10. Hayakawa, A., Goto, T., Mimoto, R., Kudo, T., Kobayashi, H. (2015), "NO formation/reduction mechanisms of ammonia/air premixed flames at various equivalence ratios and pressures," *Mechanical Engineering Journal*, Vol. 2, No. 14-00402.

SIMULATION ANALYSIS OF NH_3 MIXED COMBUSTION IN CLINKER MANUFACTURING PROCESS

Tatsuro Izumi[1], Hiroki Kujiraoka[1], Yuya Yoshizuru[1], Takeshi Suemasu[1], Makoto Ueda[1], Toyoaki Niki[1], Takayasu Itou[1], Masayuki Nishio[1], Ryuichi Murai[2] and Fumiteru Akamatsu[2],
(1)UBE Industries, LTD., UBE, Japan,
(2)Mechanical Engineering, Osaka University, Osaka, Japan

Introduction

Recent years, the action for the low-carbon society becomes active all over the world. NH_3 has potential to become the free-carbon energy source. In SIP project, that Cabinet Office started, we work on the technology development applying NH_3 to the field of industrial furnace (SIP: Strategic Innovation Promotion Program).

In this study, we evaluated the effect of NH_3 use in the cement clinker manufacturing process. Cement manufacturing is one of the fields of industry to exhaust large amounts of CO_2. As past works[1, 2)], we studied for the reduction of heat consumption rate, troubleshooting and so on by using kiln operation analysis tool, KilnSimu[3)]. This time, we applied the KilnSimu analysis technology to this work.

For some parameters using in KilnSimu, we referred to the actual plant data. About the clinker reaction kinetics under the NH_3 combustion atmosphere, we gave the element experiment. And for the initial boundary condition of gas temperature distribution, we used the combustion calculation results by CFD.

By KilnSimu analysis, we were able to evaluate the quantitative influence at using NH_3 for the actual operation. When using NH_3, the possibility of the influence on heat consumption rate was predicted. We also report the improvement plan about this.

Process Simulation

Simulator

We used KilnSimu as software of rotary kiln operation analysis. KilnSimu has several functions specialized in kiln operation analysis in addition to functions of general chemical process simulator. For example, estimation of residence time in kiln and heat flow calculation reflecting the kiln structure. **Fig. 1** shows a part of theoretical model of KilnSimu. This simulator is one-dimensional model that does gas phase / solid phase equilibrium calculation while dividing the kiln in the axial direction and calculating heat balance for each divided part.

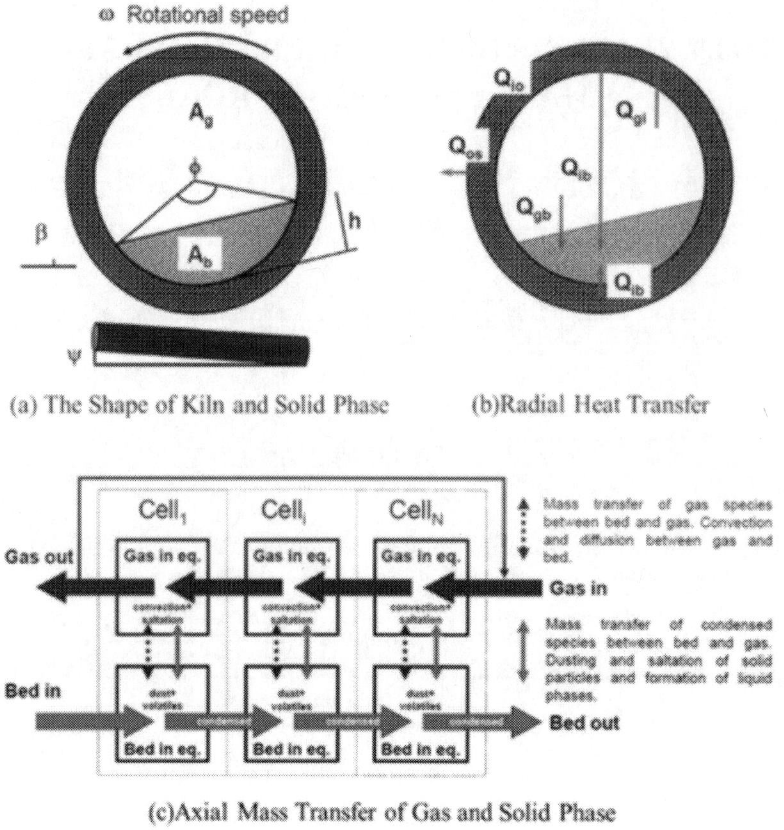

(a) The Shape of Kiln and Solid Phase (b)Radial Heat Transfer

(c)Axial Mass Transfer of Gas and Solid Phase

Figure 1. Theoretical Model of KilnSimu

Model

As the object of NH_3 mixed combustion operation analysis, we selected the Ube plant kiln (length : 87m, outer diameter : 4.8m, production volume : 200t/h) that is one of our own kilns. **Fig. 2** shows the KilnSimu model of Ube plant kiln. Like a general process simulator, KilnSimu has GUI, and it is possible to create process models by connecting blocks such as kilns and cyclones with streams. In this report, the analysis target is limited to the kiln in order to check the influence of NH_3 mixed combustion in the kiln. Thermodynamic data used in KilnSimu was data of FactSage[4)], thermodynamic equilibrium calculation software.

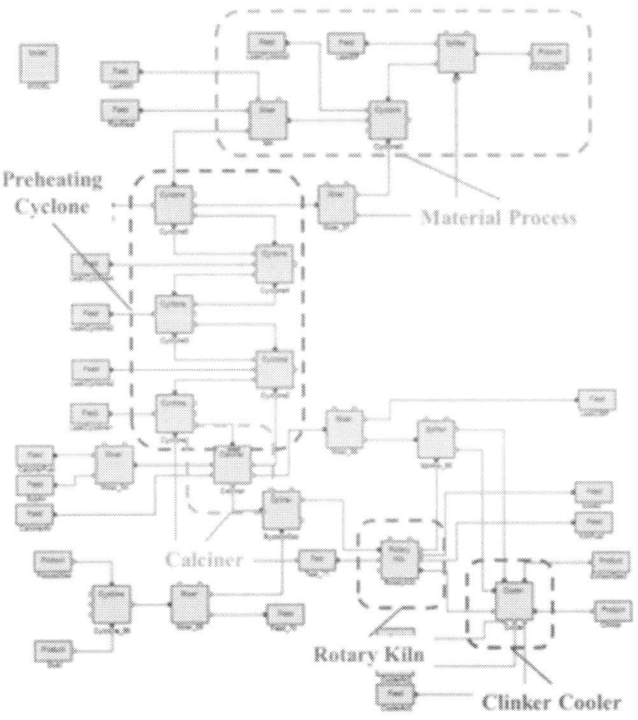

Figure 2. KilnSimu Model of Analysis Target Plant

Results and Discussion

Transition of Temperature and Materials in Rotary Kiln and Cement Clinker Composition

Fig. 3 shows the results of temperature and materials transitions in the kiln. Solid lines show the results of coal only combustion conditions, and dotted lines show the results of coal/NH_3 mixed combustion conditions. Mixed combustion rate of NH_3 is 30%. The horizontal axis of the graph indicates the position in the longitudinal direction of the kiln, and the raw material is fed on the origin side and moves in the positive direction. Calculation results under the coal only combustion conditions were obtained by inputting actual operation conditions. And it is confirmed that the temperature, gas composition and clinker composition after combustion almost agree with actual data. Calculation in the same way under NH_3 mixing conditions showed that the firing reaction for producing C3S ($3CaO \cdot SiO_2$, $2CaO \cdot SiO_2 + CaO \rightarrow 3CaO \cdot SiO_2$) did not proceed and as result the reaction consumption of CaO in raw materials was insufficient and it was shown that CaO remained to the kiln exit.

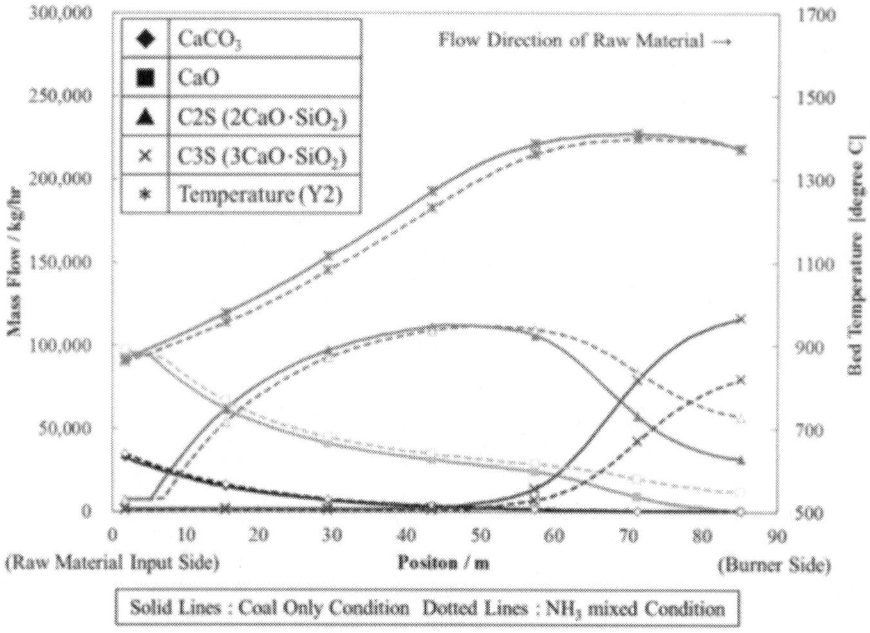

Figure 3. Calculation Results of Temperature and Materials Transitions in the Kiln

Table 1 shows the results of clinker composition at products exit. In the coal only combustion condition, C3S and f-CaO (an indicator of unreacted raw material concentration) were 59% and 0.3%, respectively. On the other hand, they were 44% and 4.7% under the NH_3 mixed combustion condition, respectively. Since f-CaO influences cement quality and strength development, it is an index to set the residual amount to less than 1%. However, it was shown that it cannot be satisfied with the 30% NH_3 mixed combustion condition examined this time. For this reason, it is thought that radiation heat transfer decreased due to a change in gas composition in the kiln and sufficient thermal energy was not supplied to the raw material. Looking at the raw material temperature in **Fig. 3**, as raw material move forward in the kiln, there is a difference in the two conditions, and in the vicinity of x = 40 – 50 m there is a difference of about 40 degree Celsius. After that, the temperature difference becomes smaller toward the kiln exit. Because it seems to be for the heat received from the flame is absorbed as reaction heat due to the progress of the C3S formation reaction (endothermal reaction), and it is not used for raising the raw material temperature. Under both conditions, C3S begins to be generated from x = 50 m or later, but the amount is much higher in the coal only combustion condition. The heat transfer from the flame to the raw material is considered to be better in the coal condition, but at the same time this is canceled out by the large amount of C3S formation reaction. And as a result, it is considered that the material temperature difference from the NH_3 mixed combustion condition becomes small.

Table 1. Calculation Results of Clinker Composition at Products Exit

	Coal Only	**NH_3 mixed**

C3S (wt%)	59.43	44.18
C2S (wt%)	16.50	27.42
C3A (Aluminate, wt%)	8.88	8.77
C4AF (Ferrite, wt%)	9.21	9.17
f-CaO (wt%)	0.32	4.68

Heat Flow in Rotary Kiln

Fig. 4 shows the heat flow rates from the combustion gas to the raw material and the inner wall of the kiln by KilnSimu. Comparing two conditions, it can be seen that the change in the heat flow rate of convection is almost constant, whereas the heat flow rate of radiation of NH_3 mixed combustion condition is greatly reduced. This is considered to be due to a change in composition of combustion gas by NH_3 mixed combustion, and a decrease in radiant intensity from gas. The emissivity of the mixed gas in KilnSimu is given by

$$\varepsilon_g = (\varepsilon_{CO2} C_{CO2} + \varepsilon_{H2O} C_{H2O})(1 - C_{SO})$$

where ε : emissivity (-), C : correction factor of emissivity (-), C_{SO} : spectral-overlap correction factor (-) and subscripts represent each component. It is calculated based on CO_2 and H_2O which are components that absorb and emit the radiation strongly, and the influence of gases with low contribution ratio such as N_2 is omitted. Approximately the emissivity based on the characteristic values described in the handbook[5] is 0.293 for coal only condition and 0.268 for NH_3 mixed condition, there is a difference.

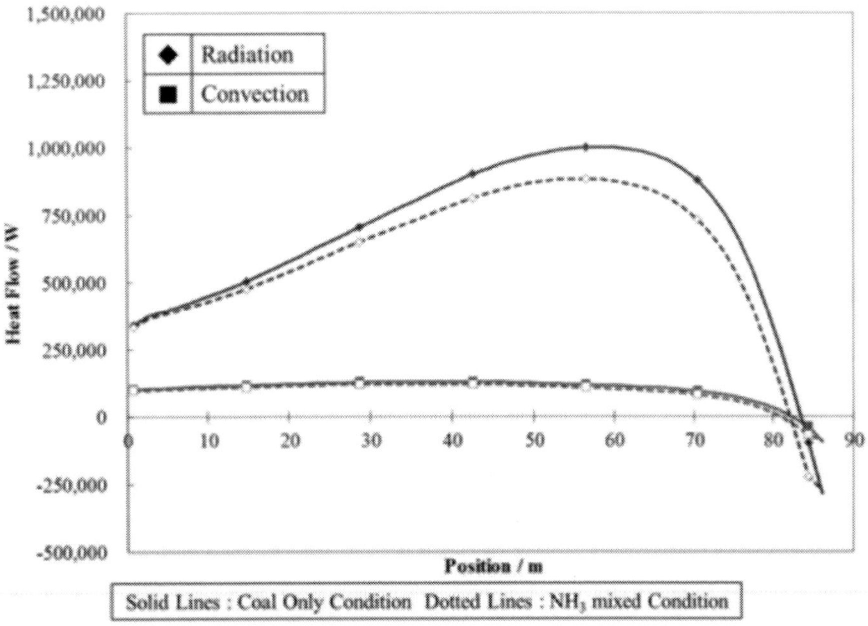

Figure 4. Heat Flow Rates from the Combustion Gas

Operational Condition for NH3 Mixed Combustion Operation

There are several ways to satisfy cement clinker quality indicators under NH_3 mixed condition. In this report, we describe prediction results of thermal efficiency improvement plans by (i) reduced production and (ii) oxygen enrichment operation.

First, **fig. 5** shows the relation between the reduced production rate and f-CaO. If we set the target value of f-CaO to be <1.0%, we found that it is necessary to reduce production by about 1.3%. We examine this result from the viewpoint of thermal balance. Calculation of the change in heat flow rate from the combustion gas at the time of coal-only and NH_3 mixed condition reduces by about 81MJ/ton. This value corresponds to about 2.1% of the thermal unit consumption of the targeted process. This means that the decrease in heat input due to mixed combustion is supplemented by the decrease in heat output due to reduced production. This time it is approximate for analyzes targeting only the kiln, but both values are balanced on the same order, which is considered to be a reasonable result.

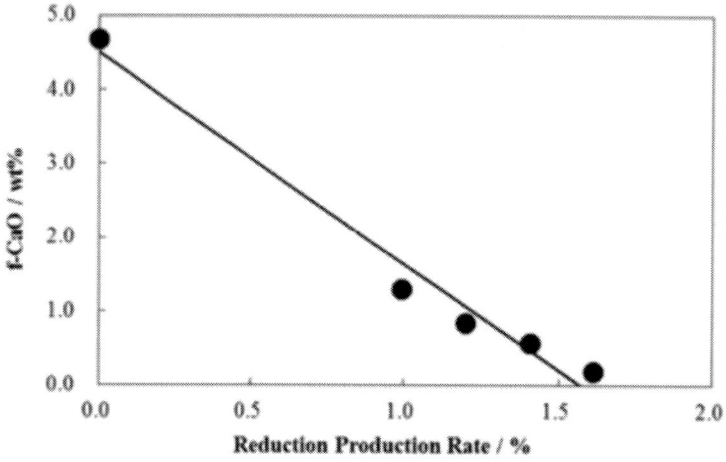

Figure 5. The Relation Between the Reduced Production Rate and f-CaO

The parts of oxygen enrichment are the primary air and secondary air of the kiln. **Fig. 6** shows the results of temperature and materials transitions. When the O_2 concentration is 23.5%, the value of f-CaO is 0.27%, about the same level as in the case of coal only combustion condition. Looking at the **fig. 6**, it is seen that the reaction is proceeding rapidly in front of the kiln exit. It is considered that this is because the gas combustion temperature increases and C2S formation reaction rate increases due to the rise in the O_2 concentration of the combustion air.

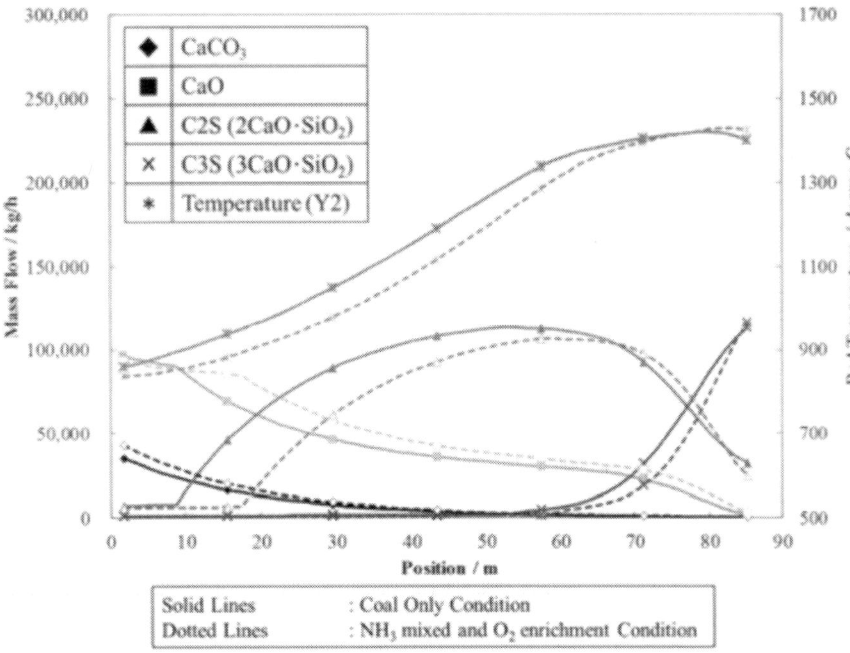

Figure 6. Calculation Results of Oxygen Enrichment Operation

Conclusions

We investigated the development technology of NH_3 mixed combustion cement kiln by simulation. The findings obtained are shown below.

- We examined the raw material temperature and the clinker composition of NH_3 mixed combustion condition. In the mixed combustion condition, the clinker forming reaction hardly proceed due to the reduction of radiation heat transfer, so it was shown that the index of f-CaO cannot be satisfied under this condition.
- As plans to improve thermal efficiency, we estimated the impact of reduced production and oxygen enrichment operation. We confirmed that both methods satisfy the index of f-CaO.

Acknowledgements

This work is supported by the Energy Carriers Program within Japan's Cross-Ministerial Strategic Innovation Promotion Program (SIP).

References

1. Ministry of Economy, Trade and Industry, Japan, (2012), "Mid-term evaluation review meeting of Innovative Fundamental Technology Development for Cement Making"
2. Ministry of Economy, Trade and Industry, Japan, (2015), "Ex-post evaluation review meeting of Innovative Fundamental Technology Development for Cement Making"
3. VTT, "KilnSimu", http://gtt.mch.rmth-aachen.de/kilnsimu
4. GTT-Technologies, "FactSage", http://gtt.mch.rmth-aachen.de/factsage

5. The Society of Chemical Engineers, Japan, (1999), "Chemical Engineering Handbook, Revised edition 6", Maruzen Publishing Co, Ltd.

Optimization of the NOx reduction condition in the combustion furnace for the combustion of "heavy-oil - NH_3 system" using CFD

Yuya Yoshizuru[1], Takeshi Suemasu[1], Masayuki Nishio[1], Ryuichi Murai[2] and Fumiteru Akamatsu[2],
1 UBE Industries,LTD., UBE, Japan
2 Mechanical Engineering, Osaka University, Osaka, Japan

Introduction

In late years the discharge of the CO_2 became the very big problem. The combustion of the fossil fuel in particular exhausts much CO_2. Our project team (SIP) is intended to reduce CO_2 by using NH_3 (10%~30%) in substitution for heavy oil. The 'SIP energy carriers' was launched in 2014 (SIP: Strategic Innovation Promotion Program). Ammonia direct combustion team was formed. We conducted a co-research program with Osaka University in this project. We performed experiment of heavy oil - NH_3 mixed combustion in the 10kW furnace. As the results, we obtained much experimental data. When we were combusted NH_3 and heavy-oil, a large quantity of NOx is exhausted. We need to conduct that out under many different conditions for NOx reduction (for example, temperature, flow rate and so on.). However, it is impossible to perform it in limited time. Therefore I found some conditions to reduce NOx using CFD. Furthermore, we introduce some conditions to optimize NOx reduction. The combustion mechanism compared the model using the detailed chemical reaction and the simplification reaction mechanism with the experiment. As a result, we became able to be combusted NH3 (30%) under a NOx condition same as heavy-oil (100%).

Fundamental experiments for heavy oil/NH_3 co-combustion in the 10kW furnace

In the 10kW industrial furnace which has been established in Osaka University, the dimensions in the furnace are as follow: H 300mm × W 300mm × L 1,200mm. The exhaust gas can measure each gas concentration of NOx, CO, CO_2 and O_2. By the thermocouple inserted from the top side, we can measure inner wall temperatures.

The fuel-oil burner become four inside ring structure. The pipe structure and put on the tip which sprays heavy oil on the most inner tube. The circumferential triple pipe part can supply a combustion air independently each and calls myself zeroth (0-Air), primary (S-Air) and second (S-Air) from the inside (Fig.1). We let the zeroth or second Air accompany by NH_3 for the purpose of investigating the influence of the supply position by this experiment and supplied it and carried out the co-combustion experiments. We did the NH_3 co-combustion rate with the substitution from 0% that became heavy-oil only to calorie base 30% who were an aim co-combustion rate. We show the experimental conditions in Table 1.

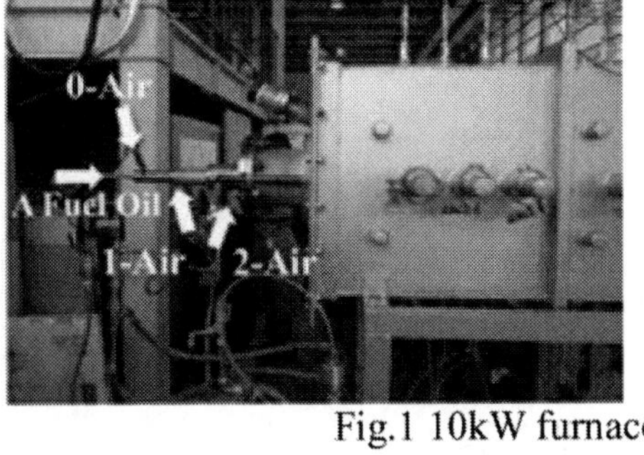

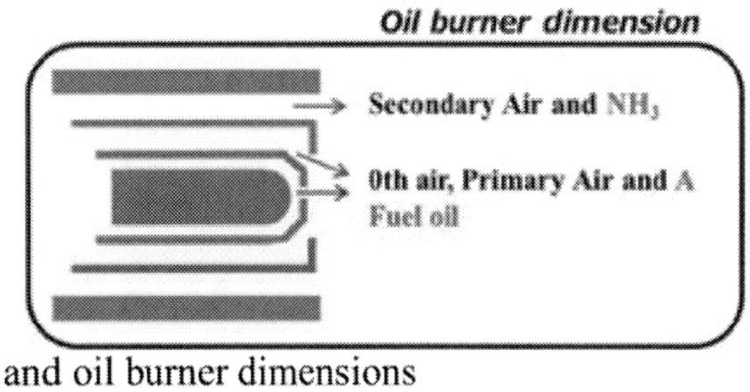

Fig.1 10kW furnace and oil burner dimensions

Table 1 Experimental conditions for λ_{all} (=1.2)

NH_3 co-combustion rate [cal%]	A Fuel Oil [L/h]	NH_3 [L/min]	0-Air [L/min]	P-air [L/min]	S-Air [L/min]
0	0.97	0.0	20.0	50.0	112.1
10	0.87	4.2	15.8	50.0	116.3
20	0.77	8.5	11.5	50.0	120.5
30	0.68	12.7	7.3	50.0	124.7

Flame properties are clearly different by the NH_3 co-combustion to show it in Fig.2. When NH_3 co-combustion rate increased in this study, the temperature distribution in the furnace decreased. N_2 increase in exhaust gas composition and a decrease in CO_2 influence it (Fig.3).

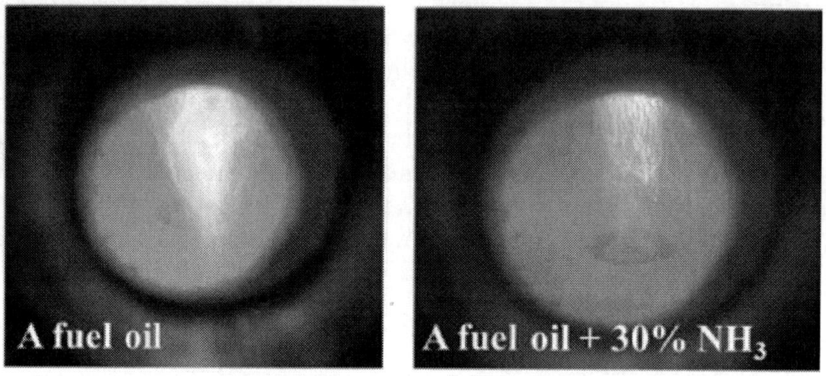

Fig.2 Combustion flame in the furnace

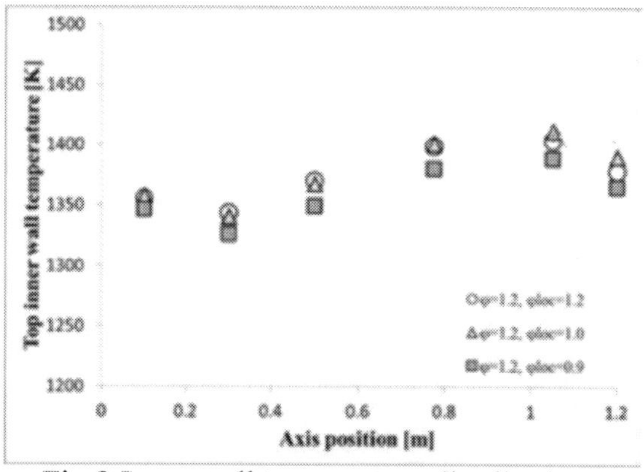

Fig.3 Inner wall temperature distributions

Here, λ_{all} is total air/theoretical air for Fuel. Otherwise, λ_{loc} is secondary air/theoretical air for NH_3. λ_{loc} is most important parameter for NOx reduction. We understood that NOx had a local minimum under the influence of λ_{loc}. We showed the results in the fig.4.

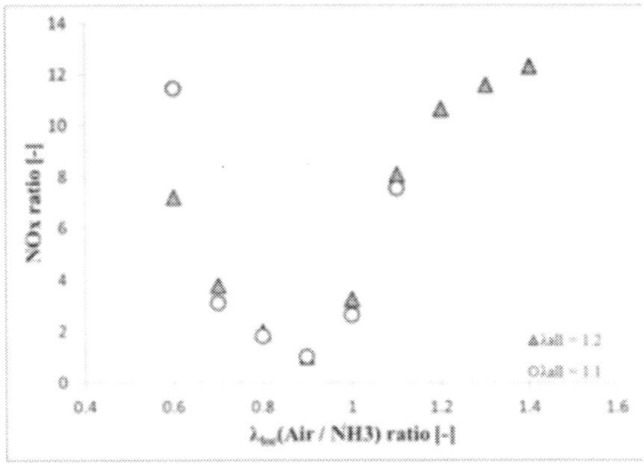

Fig.4 Experimental results for some λ_{loc}

Numerical simulation method for A Fuel Oil and NH_3 co-combustion

In order to understand the A fuel oil and NH_3 co-combustion mechanism, numerical simulations have been conducted. We solved a flow, temperature, radiation, chemical species and droplet particles to calculate combustion field. We explain a turbulence combustion model and the reaction mechanism that we used as follows.

1) Turbulence combustion model (EDC)

When it uses a general overall reactions, it is almost impossible to analyze NOx. Because some slow reactions exists, it is difficult to solve the reaction term. By the direct integration for fine scales, it is considered chemical reaction and turbulence mixing at the same time. This method is practical in the analysis of the industrial furnace.

Where species transport equations are given by,

$$\frac{\partial}{\partial t}(\rho Y_i) + \boldsymbol{\nabla}\cdot(\rho \boldsymbol{v} Y_i) = -\boldsymbol{\nabla}\cdot \boldsymbol{J}_i + R_i. \tag{1}$$

In addintions, mass fraction of chemical species are given by,

$$Y_i^{*} = Yi + \int_0^{\tau^*} \frac{M_{\mathrm{w},i}\sum_{r=1}^{N_{\mathrm{R}}} R_{i,r}}{\rho}\, dt\ . \tag{2}$$

Chemical reactions are given by,

$$\textstyle\sum_{i=1}^{N} v'_{i,r}\mathcal{M}_i \overset{k_{\mathrm{f},r}/k_{\mathrm{b},r}}{\Longleftrightarrow} \sum_{i=1}^{N} v''_{i,r}\mathcal{M}_i. \tag{3}$$

Each species net reaction rates are given by,

$$R_i = \frac{\rho(\xi^*)^2}{\tau^*[1-(\xi^*)^3]}\left(Y_i^{*} - Y_i\right), \xi^* = C_\xi\left(\frac{\nu\varepsilon}{\kappa^2}\right)^{\frac{1}{4}}, \tau^* = C_\tau\left(\frac{\nu}{\varepsilon}\right)^{\frac{1}{2}}. \tag{4}$$

Here, C_ξ and C_τ are each 2.1377 and 0.4082. These governing equations (included N-S eq., Enthalpy, DPM and Radiation) and EDC model were discretized by the control volume method and solved by AMG method. All physical properties were taken account in the dependence of the temperature.

2) Simplify reaction mechanism model (by Chemkin)

A fuel oil is very complex component. It seems that detailed reaction model is difficult. We used Model Fuel Library (MFL) by ANSYS Chemkin. MFL shows good agreement for fuel combustion. Therefore, we established the virtual fuel (for A Fuel Oil). Otherwise, there are some kind of NH_3 reaction mechanism for high temperature. But in either case, it is unfinished because a reaction pass of the reaction mechanism is uncertain. We calculated using the reaction mechanism of the fig.5 for NH_3. I think that the reaction of N_2O and the low temperature reaction paths are important to NOx reduction.

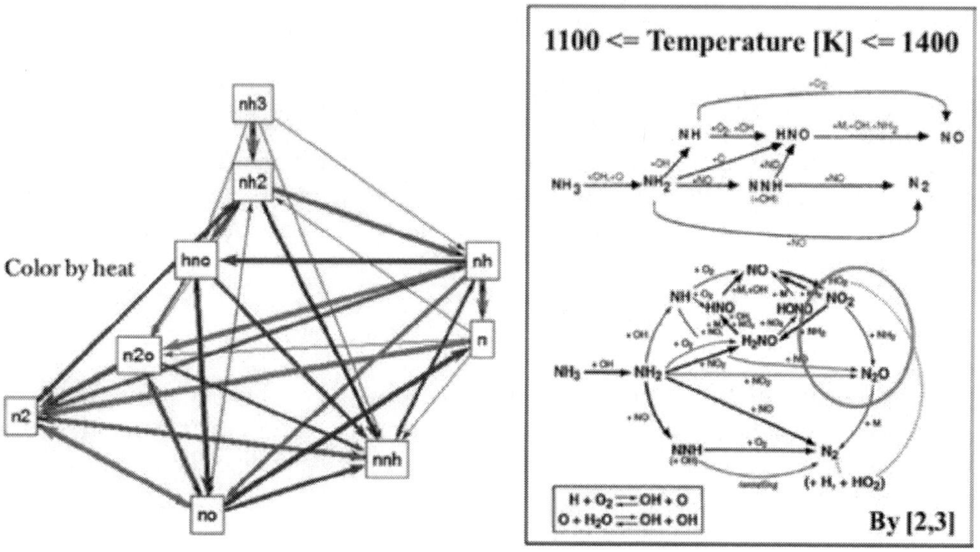

Fig.5 Reaction paths by NH_3

Results and discussion

Figure 6-7 is the contour of temperature, NH_3, NO and N_2O for λ=1.2, λ_{loc}=1.2. Please observe temperature distribution. This side of the combustion zone becomes the black zone. The flame shape is wide. NH_3 fuel is slow combustion. I was able to confirm that most of the NOx were of fuel NOx origin.

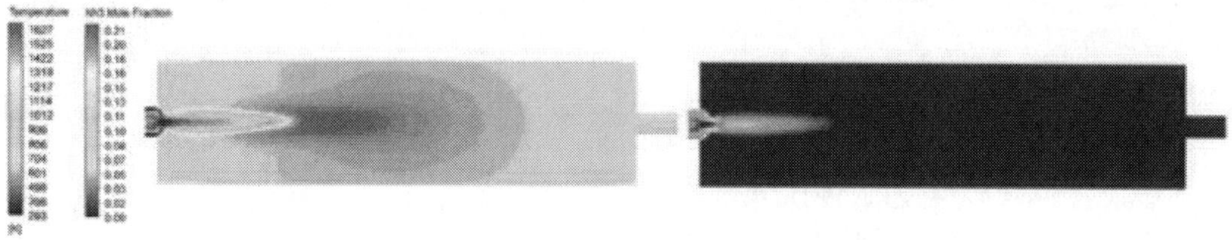

Fig.6 The contour of temperature and NH_3 for λ=1.2, λ_{loc}=1.2

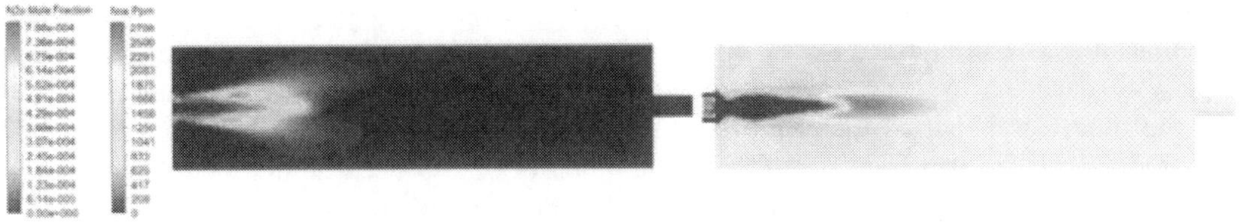

Fig.7 The contour of NO and N_2O for λ=1.2, λ_{loc}=1.2

A difference became less than the experimental results in the NH_3 reaction mechanism which we used in this study. I am inspecting how there is influence of the reaction mechanism that I simplified (Fig.8). But reaction paths are some difference. Both NNH and NH are slightly different. Therefore, NO concentration was changed behind the temperature peak.

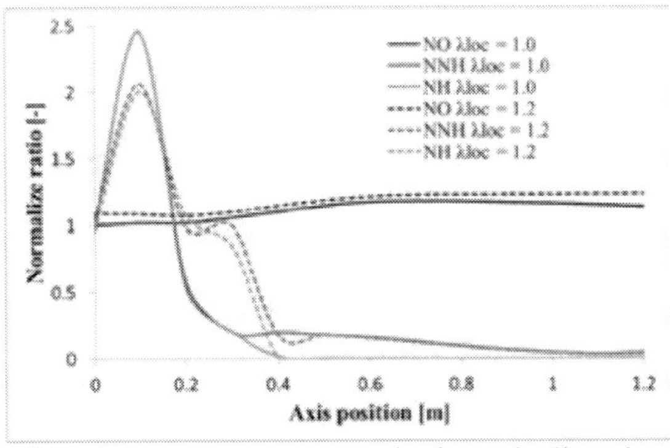

Fig.8 The concentration ratio for axis direction

Figure 9 is inner wall temperature distribution in furnace and NO concentration ratio of exhaust gas. The result of temperature distribution tendency corresponded with our experiments. The NO ratio for λ_{loc} was overestimated for the experimental results. But the quantity of NO became the tendency to decrease in the same way as our experiment results. I am calculating other cases and am going to report it after the next time.

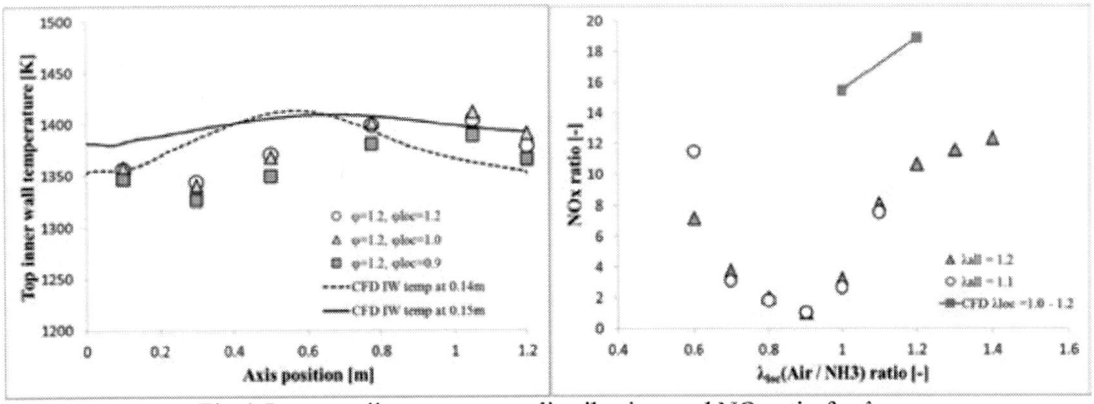

Fig.9 Inner wall temperature distribution and NO ratio for λ_{loc}

Conclusions

In this study, we conducted heavy oil/NH_3 co-combustion experiment and numerical simulation in the 10kW furnace. As the result, we became able to reduce NOx by controlling air to convey with NH_3. On the other hand, I tried that I reproduced an experiment by CFD. The NO ratio for λ_{loc} was overestimated for the experimental results. But the quantity of NO became the tendency to decrease in the same way as our experiment results. I revise calculation precision and reaction mechanism and am going to challenge it again.

Acknowledgements

This work was supported by Council for Science, Technology and Innovation(CSTI), Cross-ministerial Strategic Innovation Promotion Program (SIP), "energy carrier"(Funding agency:JST).

Nomenclature

J_i	$:= -\left(\rho D_{m,i} + \frac{\mu_t}{Sc_t}\right) \nabla Y_i$ Flux of the species	[kg m^{-2} s^{-1}]
$k_{f,b,r}$	: Rate constant for the forward and backward	[s^{-1}]
M_i	: Molecular mass of species	[g mol^{-1}]
T	: Temperature	[K]
$\boldsymbol{v}$	: Velocity vector	[m s^{-1}]
$\nu_i', ''$	: Stoichiometric coefficient of reaction	[-]
Y_i	: Mass fraction of component *i*	[-]
ρ	: Density of gas	[kg m^{-3}]

References

1. Akihiro Hayakawa, Takashi Goto, Rentaro Mimoto, Taku Kudo, Hideaki Kobayashi (2015), "NO formation/reduction mechanisms of ammonia/air premixed flames at various equivalence ratios and pressures," *Bulletin of the JSME*, Vol.2, NO.1
2. Bjørn F. Magnussen (1981), "On the Structure of Turbulence and a Generalized Eddy Dissipation Concept for Chemical Reaction in Turbulent Flow," *19th AIAA Aerospace Science Meeting*, Jan. 12-15 1981, St.Louis, Missouri
3. James A. Miller, Craig T. Bowman (1989), "Mechanism and Modeling of Nitrogen Chemistry in Combustion," *Prog.Energy.Combust Sci*, Vol.15, pp.287-338
4. James A. Miller, Peter Glarborg (1996), "Modelling the Formation of N_2O and NO_2 in the Thermal De-NOx Process," *Gas Phase Chemical Reaction Systems*, pp.318-333
5. Z.Tian, Yuyang Li, Lidong Zhang, Peter Glarborg, Fei Qi (2009), "An experimental and kinetic modeling study of premixed $NH_3/CH_4/O_2/Ar$ flames at low pressure," *combustion and Flame*, 156, pp.1413-1426

540693 Ignition of an Aqueous Ammonia/Ammonium Nitrate Fuel

Abstract:

Ignition of an aqueous ammonia/ammonium nitrate fuel

Bar Mosevitzky, Gennady E. Shter, Gideon S. Grader*

*Corresponding author: grader@technion.ac.il

Wolfson Department of Chemical Engineering, Nancy and Stephen Grand Technion Energy Program, Technion-Israel Institute of Technology, Haifa 3200003, Israel

To achieve a truly renewable energy market, the intermittent power generation of sources such as solar and wind must be overcome. Renewable ammonia can be synthesized using these sources to be used as a long-term energy storage medium. For this reason, the use of ammonia as a synthetic fuel has garnered significant attention in recent years. Aqueous AAN (ammonia/ammonium nitrate) is a carbon-free ammonia based monofuel suitable for energy storage applications. This fuel is safe to store and transport, and its combustion products consist mainly of water and nitrogen. Effective use of this fuel requires an in-depth understanding of the reaction pathways dominating its ignition.

In this work, the accumulated results from experiments conducted to test the effects of water content (Fig. 1a), equivalence ratio (Fig. 1b) and diluent pressure (Fig. 1c) on AAN ignition will be reviewed. The use of simulations to reproduce these results will be evaluated, and the data generated by rate-of-production and sensitivity analyses will be reviewed. Finally, the reaction pathways involved in AAN ignition and their relation to its water content, equivalence ratio and diluent pressure will be presented, and the implications of the rate-determining steps for AAN ignition will be discussed.

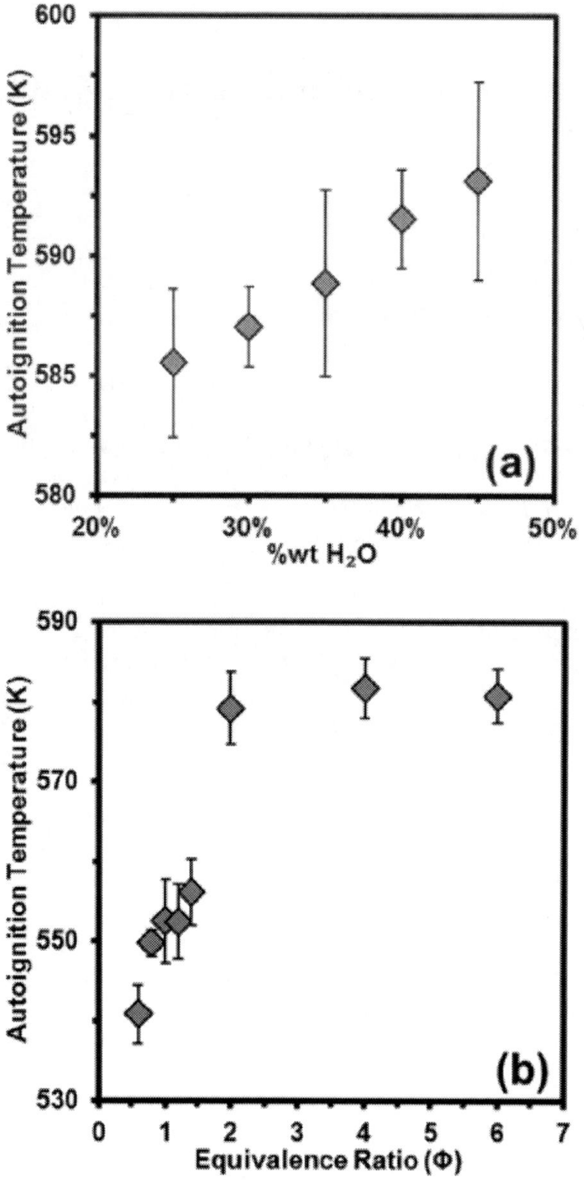

600
595
590
585
580
Autoignition Temperature (K)
(a)
20%
30%
40%
50%
%wt H2O
590
570
550
530
Autoignition Temperature (K)
(b)
0
1
2
3
4
5
6
7
Equivalence Ratio (Φ)

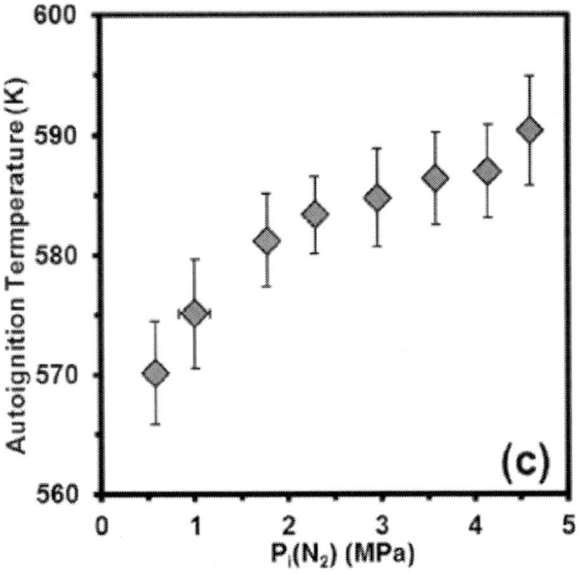

Figure 1. The effects of: (a) water content, (b) equivalence ratio and (c) diluent pressure on the Autoignition Temperature (AIT) of AAN.

541131 Improved Method of Using Hydrogen and Ammonia Fuels for an Internal Combustion Engine

Abstract:

A tractor mounted internal combustion engine is fueled by Hydrogen or a combination of Hydrogen and Ammonia.

Developments of an improved method of fuel injection and ignition control. Hydrogen is port injected in the intake manifold, and liquid ammonia is injected in the throttle body. A dual fuel ECU, Engine Control Unit, controls the fuel mixtures and the firing of multiple coils for ignition.

The paper will address significant engine performance improvements and the resulting fuel consumption and engine emissions levels.

Link to project website: www.solarhydrogensystem.com

540336 Auto-Ignition Kinetics of Ammonia at Intermediate Temperatures and High Pressures

Abstract:

The anxiety over global greenhouse gas emissions has intensified the demand for the development and use of CO_2-neutral energy technologies. Ammonia is now attracting attention as a carbon-free energy carrier, because it has good energy density (22.5 MJ/kg) and can be easily liquefied (about 10 bar at 298 K). In addition, ammonia is produced according to the Haber-Bosch process, which makes it one of the most widely-produced inorganic chemical in the world. It could also be produced with renewable energy sources such as wind and solar energy using P2X technology.

As a potential fuel for applications in gas turbines and gas engines, ammonia is less reactive than most hydrocarbons and its ignition and combustion characteristics are not yet well understood. A major part of the previous research has focused on the ammonia oxidation at high temperatures or low pressures [1-3], while ignition measurements for pure ammonia or ammonia mixed with other gaseous fuels (such as hydrogen or methane) at high pressures and low-to-intermediate temperature is rare.

Rapid compression machines (RCMs) are regarded as an important experimental apparatus for investigating auto-ignition behavior at low-to-intermediate temperature conditions, which are quite relevant to the application in internal combustion engines and gas turbines [4,5].

In this study, autoignition properties of NH_3/O_2 and $NH_3/H_2/O_2$ mixtures have been studied in a RCM at pressures from 20 to 60 bar, temperatures from 950 to 1150 K, and at equivalence ratios from 0.5 to 2. The effect of hydrogen-ammonia ratio in fuel has been also investigated. It is observed in the experiments that a higher H_2 mole fraction in fuel improves the reactivity of the mixture. When the mixtures contain 20% H_2 in fuel (xNH_3:xH_2=0.8:0.2), fuel-richer mixtures have shorter ignition delay times, while for the mixtures with 1% H_2 the equivalence ratio dependence is reverse. With 5% H_2 in fuel, the stoichiometric mixture presents the shortest ignition delay time. In the mixtures without hydrogen, leaner mixtures show higher reactivity. Numerical simulations were performed based on the literature mechanism from Klippenstein et al. [6] and Glarborg et al.[7]. The model predicts well the ignition delay times for the mixtures with 20%, 1% H_2 and pure ammonia. For the mixtures with 5% H_2 the model underestimates the ignition delay time. Kinetic analysis further demonstrates that the reactions involving H_2NO and the two channels of the reaction NH_2 + NO are dominant for the ammonia auto-ignition in the investigated temperature range, and the reactions involving H_2O_2 are important for ammonia-hydrogen blends.

[1] Mathieu,O., Petersen,EL., Combust Flame 2015 ;162:554-70.
[2] Nakamura, H., Hasegawa, S., Proc Combust Inst 2017; 36: 4217-26.
[3] Brackmann, C., Alekseev, V.A., Zhou, B., Nordström, E., Bengtsson, P.E., Li, Z., et al., Combust Flame 2016; 163: 370-81.
[4] Sung, CJ., Curran, H.J., Prog Energy Combust. Sci 2014; 44: 1-18.
[5] Goldsborough, S.S., Hochgreb, S., Vanhove, G., Wooldridge, M.S., Curran, H.J., Sung, CJ., Prog Energy Combust. Sci 2017; 63: 1-78.
[6] Klippenstein, S.J., Harding, L.B., Glarborg, P., Miller, J.A., Combust Flame 2011; 158: 774-89.
[7] Glarborg, P., Miller, J.A., Ruscic, B., Klippenstein, S.J., Prog Energy Combust. Sci 2018; 67: 31-68.

Experimental and Computational study for Reduction of NOx emissions in the ammonia/methane co-combustion in a 10 kW furnace

Ryuichi Murai*1, Ryohei Omori[1], Yuya Yoshizuru[2], Takahiro Kitano[1], Hidetaka Higashino[1], Noriaki Nakatsuka[1], Jun Hayashi[3] and Fumiteru Akamatsu[1]

[1]Department of Mechanical Engineering, Osaka University
2-1 Yamadaoka, Suita, Osaka 565-0871, Japan

[2]Ube Industrials. Ltd.,
1978-10 Oaza-ogushi, Ube, Yamaguchi 7558633, Japan

[3]Graduate School of Energy Science, Kyoto, University
Yoshida Honmachi, Sakyo-ku, Kyoto 6068501, Japan

Introduction

There are severe issues on increasing amount of carbon dioxide (CO_2) emission in the world. Many studies are devoted to alternative fuels. One of promising candidates is the utilization of ammonia which is zero emission of CO_2, a hydrogen energy carrier, and also can be burned directly as a fuel. For direct combustion of ammonia in industrial furnaces, there were two issues which were weaker radiative heat flux and a huge amount of NOx emission compared with the combustion of methane. We already have reported [1] the solution of the former issue by using the oxygen enriched combustion. The objective of this research is to study the reduction mechanism of NOx emissions in the ammonia/ methane co-combustion in an industrial furnace both experimentally and numerically.

Experiments

Experimentally we measured the radiation spectra and the total radiative thermal flux under the condition of the ammonia fuel burned in a 10 kW furnace with a coaxial jet flame and additional two

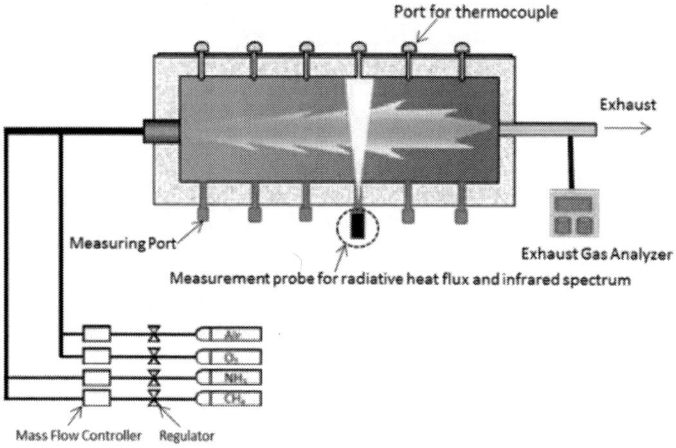

Figure1 Experimental Apparatus

oxidizer inlets for the staging combustion. Figure1 shows an experimental schematic diagram. A furnace is 1400 mm long and its inner wall is surrounded with the adiabatic material. On a top of the furnace there are six ports equipped with thermocouples which are positioned at Z=100 mm, 200 mm, 300 mm, 500 mm, 750 mm, 1000 mm from the exit surface of a burner, where we define Z=0 mm. On a bottom of a furnace, there also are six ports for measuring heat flux and emission spectrum from the inner wall and the flame. But those two data cannot get at the same time. For the thermal emission spectrum measurement, we installed the spectrum detector instead of the heat flux detector. Therefore, we can compare heat flux data with spectrum ones at the same position

Results

The spectrum measurement results show that the amount of NOx emission was in reverse proportion with the intensity of N_2O spectrum in the downstream of the reaction zone in the furnace. This indicates that N_2O, which is one of main intermediate spices of NH_3, reacts with NOx as a reduction reactant to nitrogen molecule.

Numerical simulations were performed to clarify these hypotheses. Firstly we carried out the simulation to overview the combustion in the whole area of a furnace, with the condition of three dimensions, eddy dissipation concept model in the turbulent combustion and GRI-Mech3.0 in the ammonia and methane reaction.

When the amount of NOx emission was smaller at the furnace exit, more N_2O existed in the middle region of the furnace. This simulation could explain the experimental data with good agreement.

Reference

[1] R. Murai et al, The radiative characteristics of NH3/N2/O2 non-premixed flame on a 10 kW test furnace, Energy Procedia, vol. 120, August 2017, P325-332

540576 A Low Pressure Membrane Based Renewable Ammonia Synthesis

Abstract:

A Low Pressure Membrane Based Renewable Ammonia Synthesis

Sarb Giddey, Danielle Kennedy, Ani Kulkarni, Gary Paul, David Alexander

Presenting authorâ€™s e-mail: sarb.giddey@csiro.au

CSIRO, Private Bag 10, Clayton South, VIC, Australia

Abstract

Ammonia is currently mostly produced by the highly energy and carbon-intensive Haberâ€"Bosch process, which requires temperatures of 450â€"500°C and pressures of up to 200 bar. The feedstock for this process is hydrogen from natural gas (NG), coal or oil, and nitrogen produced from air by cryogenic route or pressure swing adsorption (PSA). The share of NG, coal and fuel oil feedstock for the global production of ammonia is 72%, 22% and 4% respectively, contributing to approximately 420 million tons of CO_2 emissions per annum, representing over 1% of global energy related emissions. The energy consumed for ammonia synthesis by Haber-Bosch process is in the 10 to 15 MWh/tonne range, depending on the type of fossil fuel feedstock used.

In an alternative route renewable hydrogen produced by an electrolyser can be fed to the Haber-Bosch reactor along with nitrogen for ammonia synthesis, and this route has been suggested to consume energy around 12 MWh/tonne of ammonia.

CSIRO has developed a metal membrane based ammonia synthesis process that uses hydrogen and nitrogen as feedstock. The materials and catalyst developed for the process allow the synthesis process at much lower pressures (~ 10 bar) at 450°C. The synthesis rates achieved are two orders of magnitude higher than with any electrochemical route. The catalyst and the metal / catalyst interfacial structure have been specifically tailored for low pressure ammonia synthesis. The low pressure membrane reactor allows direct coupling to an electrolyser and air separation unit (ASU) operated by a renewable source, thus promising over 25% reduction in the energy input, substantial capital savings on reactors compared to conventional Haber-Bosch process, and allows distributed or centralised ammonia production.

This presentation will provide details and current status of the technology in terms of the catalyst and interfacial structures developed, and the reactor design to achieve high ammonia synthesis and conversion rates. The presentation will also discuss the merits of this technology compared to the conventional and other emerging routes.

535869 Identifying the Prospects of Electrochemical Ammonia Synthesis on Mxenes Using First Principles Calculations

Abstract:

Electrochemical synthesis of ammonia is a major challenge aimed at making production of ammonia sustainable. Currently ruthenium is the transition metal of choice for catalyzing the industrial Haber-Bosch process. However, electrochemical ammonia synthesis on ruthenium suffers from high overpotential and the competing hydrogen evolution reaction. Recently layered transition metals carbides and nitrides (MXenes) have been identified as a potential material class for ammonia synthesis. MXenes are particularly interesting owing to the high degree of tunability in surface chemistry due to the transition metal choice, interlayer distance, number of layers in the material, and surface termination. These choices affect the electron density of the surface and hence the binding strength of MXenes with key intermediates. In this work, we use density functional theory (DFT) to compute adsorption free energies of relevant intermediates to identify MXenes that are promising for ammonia synthesis. Using uncertainty quantification capabilities within the Bayesian error estimation functional (BEEF), we also compute the probability density functions for catalytic activity predictions. We obtain free energy diagrams and scaling relations and finally report prediction confidence values on the limiting potential and insights into the prospects of using MXenes for nitrogen fixation.

527378 Highly-Selective Electrochemical Reduction of Dinitrogen to Ammonia at Ambient Temperature and Pressure

Abstract:

Catalytic conversion of dinitrogen (N_2) into ammonia under ambient conditions represents one of the Holy Grails in catalysis and surface science. As a potential alternative to the Haber-Bosch process, electrochemical reduction of N_2 to NH_3 is attractive owing to its renewability and flexibility, as well as sustainability for producing and storing value-added chemicals from the abundant feedstock of water and nitrogen on earth. However, owning to the kinetically complex and energetically challenging N_2 reduction reaction (NRR) process, NRR electrocatalysts with high catalytic activity and high selectivity are rare. In this contribution, as a proof-of-concept, we demonstrate that both the NH_3 yield and NH_3 faradaic efficiency (FE) at ambient conditons can be improved by modification of the hematite nanostructure surface. The hematite electrocatalysts were fabricated with different numbers of surface oxygen vacancies. A higher concentration of surface oxygen vacancies renders improved performance toward NRR. A maximum ammonia production rate of 0.46 μg h^{-1} cm^{-2} and a maximun FE of 6.0% have been achieved at −0.9 V *vs* Ag/AgCl in 0.1 M KOH electrolyte. This is the highest FE reported so far among non-precious metal based catalysts using a polymer-electrolyte-membrane type cell and higher than those of the precious metal based catalysts (Ru/C and Pt/C) under comparable reaction conditions. This work suggests the important role of surface states in transition metal oxides for promoting electrocatalytic reduction of N_2 to NH_3 under ambient conditions and may spur interest toward rational structure design of efficient electrocatalysts for NRR with enhanced activity and selectivity.

520287 Electrochemical Synthesis of Ammonia Using Metal Nitride Catalsyts

Abstract:

With the development of the Haber process and the subsequent work done by Bosch, ammonia production become an industrially and economically viable way to fix nitrogen. This helped increase the global population and estimates put it at about 40% of the global populationâ€™s food comes from ammonia made by the Haber-Bosch process[1]. However, the Haber-Bosch process is an energy intensive process requiring high pressure (15-30 MPa) and relatively high temperature (430 Â°C â€“ 480 Â°C) and is highly centralized with only about 13 companies and about 29 plants[2, 3]. Renewable energy resources offer a possible alternative way to fix nitrogen at low temperature and low pressure to produce ammonia in a decentralized way. High temperature solid proton conductors have been used to produce ammonia selectively, but the high temperature can degrade the ammonia[5] and the required heating makes distributed production difficult. Low temperature polymer electrolyte membranes can be used which might reduce the overall energy input. The most prevalent polymer electrolyte is an acidic proton exchange membrane (PEM). Platinum group metals (PGMs) are the standard option for electrochemical testing, especially in PEM setups. However, they generally have low Faradaic efficiencies (FE) of less than 1% toward the electrochemical nitrogen reduction reaction (ENRR) with the rest going toward the hydrogen evolution reaction (HER). We examined other non-PGM catalysts for ammonia production, focusing on metal nitrides because metal nitrides have been predicted, computationally, to be more effective catalysts for ENRR, however, experimental verification has been lacking. We previously have shown that a transition metal nitride (Cr_2N) is an active and stable ENRR catalyst in a PEM electrolyzers. In this work, we compare the activity of a variety of monometallic metal nitrides, and demonstrate how the composition of mixed metal nitrides, specifically chromium nitride and vanadium nitride. affects the overall ENRR stability and activity.

References:

1. Erisman, J.W., et al., Nature Geosci, **1**(10): p. 636.(2008).
2. *Mineral Commodity Summaries 2015*, U.S.G. Survey, 2015: Reston, Virginia. p. 196.
3. Marnellos, G. and M. Stoukides, Science, **282**(5386): p. 98.(1998).
4. Kordali, V., G. Kyriacou, and C. Lambrou, Chem. Commun., (17): p. 1673.(2000).
5. Amar, I.A., et al., J Solid State Electrochem, **15**(9): p. 1845.(2011).
6. Lan, R. and S. Tao, RSC Advances, **3**(39): p. 18016.(2013).
7. Sheng, W., H.A. Gasteiger, and Y. Shao-Horn, J. Electrochem. Soc., **157**(11): p. B1529.(2010).

529549 Electrochemical Nitrogen Reduction Reaction on Transition Metal Nitride Nanoparticles in Proton Exchange Membrane Electrolyzers

Abstract:

Transition metal nitride nanoparticles are synthesized and utilized as catalysts for electrochemical nitrogen reduction reaction (ENRR) to produce ammonia in a proton exchange membrane electrolyzer (PEMEL). The catalysts show an average ENRR rate and Faradaic efficiency (FE) of 3.3×10^{-10} mol s^{-1} cm^{-2} (6.6×10^{-10} mol s^{-1} mg^{-1}) and 5.95% at −0.1 V within 1 h, respectively. Both the ENRR rate and FE are approximately two orders of magnitude higher than those of noble metal catalysts. Time-dependent results suggest that the catalytic activity of transition metal nitride nanoparticles is stable at −0.1 V, with the catalytic activity decreasing by only 28% during 4 h test. *Ex situ* X-ray photoelectron spectroscopy (XPS) and *operando* electrochemical and synchrotron-based X-ray absorption spectroscopic (XAS) analysis reveal that transition metal oxynitride is the active phase for nitrogen reduction.

535806 DFT Analysis of Elementary N_2 Electro-Reduction Kinetics on Transition Metal Surfaces

Abstract:

Ammonia is currently produced through the catalytic Haber Bosch process (HB) at temperatures of about 300 to 500 ^{O}C and pressure of about 200-300 atm. In a future with plentiful renewable electricity from distributed sources, an electro-chemical system to produce ammonia could efficiently generate ammonia on site and on demand. Possible heterogeneous catalysts for electro-chemical nitrogen reduction are currently marred by the poor rate and selectivity due difficulty in activating the strong N-N bond and to the competing hydrogen evolution reaction (HER), resulting in infeasible faradaic efficiency. To develop more selective and active catalysts, better understanding of the mechanistic and kinetic aspects of nitrogen electro-reduction is essential. We use density functional theory (DFT) methods to examine the elementary kinetics of the possible associative mechanisms for electro-reduction of N_2. A comprehensive electro-reduction mechanism with elementary activation barriers for each step will be presented on iron (Fe) surfaces. Key step barriers are also evaluated across a series of late transition metal catalysts to evaluate the reliability of scaling and Bronsted-Evans-Polanyi relationships to predict catalyst performance. We compare the â€œkinetic over-potentialâ€ to the thermodynamic over potential to emphasize the need to explicitly evaluate barriers in predicting catalyst performance. We then examine how near-surface additives can act as shuttling agents to alter the kinetic barriers of the proton coupled electron transfer, acting as co-catalysts to improve activity and selectivity.

529736 Enhanced Electrochemical Ammonia Production Via Peptide-Bound Metal

Abstract:

Approximately half of the people on the planet are alive because of synthetically produced ammonia. However, due to the fossil fuels used in the current ammonia synthesis process, its production contributes a significant amount to the worldâ€™s greenhouse gas emissions. Haber-Bosch synthesis, which is the most widely used method of producing synthetic ammonia today, requires high temperatures (400-500 Â°C) and pressures (150-200 atm). This process is also energy intensive, consuming approximately 2% of worldwide energy. By taking an electrochemically-based approach to ammonia synthesis, those harsh conditions and emissions can be eliminated. However, current catalysts are not selective for the desired reaction.

We hypothesize that 3D surface modifications can be utilized to overcome these selective limitations and create catalysts which mimic the selectivity of the nitrogenase enzyme: an enzyme that catalyzes the reduction of nitrogen to ammonia at mild temperatures and pressures. Therefore, in this study, we show that a peptide sequence, when bound to iron(iii) oxide metal nanoparticles, facilitates electrochemical ammonia with relatively high efficiency. These recent results were obtained in an alkaline, solid-state electrochemical cell, yielding an electrochemical ammonia production rate and faradaic efficiency that is >10x higher than the catalyst without bound peptide. These results are corroborated in an alkaline liquid-based cell.

This discussion covers the characterization of the peptide and its interactions with iron(iii) oxide. The goal of this study is to elucidate the peptideâ€™s effects on material properties and how they might lead to increased ammonia production. Techniques and instrumentations used to investigate our system involve a quartz crystal microbalance with dissipation (QCM-D) with humidity and electrochemistry modules, x-ray photoelectron spectroscopy (XPS), BET gas adsorption, circular dichroism (CD), and various electrochemical techniques. The insights revealed in this work will eventually be used to direct material optimization for efficient electrochemical ammonia production.

540566 New Insights into Electrocatalysis of Nitrogen Reduction to Ammonia

Abstract:

Ammonia was electrochemically produced from nitrogen and water using a ruthenium–platinum (RuPt) alloy catalyst cathode and a nickel anode at ambient pressure and room temperature. The rate of ammonia formation was 5.1×10^{-9} gNH_3 s^{-1} cm^{-2} with a 13.2% faradaic efficiency at an applied potential of 0.123 V vs. RHE; it reached 1.08×10^{-8} gNH_3 s^{-1} cm^{-2} at 0.023 V. Ammonia production was investigated under selected potentials and temperatures. Real-time direct electrochemical mass spectrometric (DEMS) analysis of the evolved gases was performed at various applied potentials. In general, the mass-to-charge ratio signals of hydrogen and ammonia were detected, and their intensities increased with increasing potentials; however, there was no trace of a hydrazine signal. Compared to metallic ruthenium and platinum catalysts, RuPt showed a synergistic effect toward electrochemical formation of ammonia due to co-catalysis.

532475 Electrochemical Reduction of Nitrogen to Ammonia over Transition Metals

Abstract:

The ability to produce ammonia in a sustainable and efficient manner has been a topic of scientific and industrial importance for many years. The Haber-Bosch process has acted as the primary process for transforming nitrogen and hydrogen gas into ammonia. This process has become unsustainable in the foreseeable future and requires a cost-effective alternative. Ammonia is a critical component of fertilizer that is vital to the agriculture industry. The electrochemical reduction of N_2 to ammonia would eliminate carbon dioxide emissions that are present in current ammonia production processes and allow for a environmentally favorable process. Although the electrochemical reduction of nitrogen to ammonia has been researched as an alternative process, the high temperature and pressure requirements have not allowed for the advancement of the field. This project further investigates the potential for electrochemical reduction of nitrogen to ammonia via transition metal electrocatalysts. The challenge is to develop not only an active, but selective catalyst for ammonia production. We utilized a compact, parallel plate electrochemical cell that allowed for the study of transition metal activity for electrochemical reduction of nitrogen at ambient temperature and pressure. The product distribution resulting from the electrochemical reduction of nitrogen and the corresponding Faradaic efficiencies have been reported for first and second row transition metals. Moreover, several operating conditions were tested and electrocatalytic stability was monitored.

540634 A Study on Electrochemical Ammonia Synthesis with Proton-Conducting Solid Oxide Electrolytic Cells Based on $La_{0.8}Sr_{0.2}Ga_{0.8}Mg_{0.2}O_{3-D}$

Abstract:

Ammonia has become one of the most important chemicals with its versatility since the Haber-Bosch process was invented. Recently, ammonia has been getting interests because of its possibility as a hydrogen carrier. Since ammonia has high energy density and carbon-free characteristics, using ammonia as a fuel of solid oxide fuel cells is advantageous. However, the Haber-Bosch process spends much electricity because of the high pressure condition, and the process consumes more than 1% of energy consumption worldwide. Therefore, the development of a new method for the ammonia production is necessary. In this study, solid oxide based electrolytic cells were fabricated to synthesize ammonia in an electrochemical way. Various metal catalysts were tested with the electrolytic cells to synthesize ammonia, and an infiltration method was introduced to enlarge triple-phase boundaries of the cells. Operating temperatures and bias were also varied to find the feasible condition for the electrochemical ammonia synthesis. The production rate was compared to the previously reported values, and it is over 1Ã—10^{-9} mol NH_3â˙™sâ˙™cm^{-2}. Also, to see the feasibility of the ammonia operation of the fuel cells, the performance of solid oxide fuel cells was tested under ammonia-fed condition and compared to hydrogen-fed condition. At the single-cell level, the performance under the ammonia-fed condition showed about 20% difference compared to the hydrogen-fed condition.

540620 Low-Pressure Electrolytic Ammonia Synthesis Via High-Temperature Polymer-Based Proton Exchange Membrane

Abstract:

The University of North Dakota Energy and Environmental Research Center (EERC) and North Dakota State University (NDSU) have developed a low-pressure electrolytic ammonia (LPEA) production process. The LPEA process uses an electrochemical cell based on an innovative polymerâ€“inorganic composite (PIC) high-temperature (300Â°C) gas-impermeable proton-exchange membrane conceptualized and partially developed by EERC and NDSU. Because of its operability at ambient pressure and quick start-up capability (versus traditional high-pressure Haber Bosch-based plants), the LPEA process offers compatibility with smaller-scale plants and intermittent operation, and a cost-effective means of monetizing (and storing) renewable energy as ammonia. EERC, NDSU, and Proton OnSite are embarking on a project to optimize the PIC membrane, optimize and assess the technical and commercial viability of the LPEA process, and develop an LPEA process commercialization plan. To be presented will be information and data regarding 1) the configuration and working principles of the PIC membrane and LPEA process, 2) how the PIC membrane and LPEA process will be optimized, and 3) plans for LPEA process demonstration and commercialization.

540746 Atmonia: Sustainable Ammonia Production Using Electrocatalysis at Ambient Temperature and Pressure

Abstract:

Density functional theory simulations have shed light on reaction mechanisms, rate limiting steps and minimum energy paths for reactions to occur, in vacuum as well as in various media. Using that, we have selected certain criteria and revealed a few metal nitride surfaces that should be efficient and selective catalysts for nitrogen reduction in water. Recently, experimental confirmation was acquired for one of the surfaces. A novel methodology was developed where electrochemical catalysis chamber was directly connected in-line with a flow injection analysis method, providing direct detection of reaction rate and catalyst current efficiency, which is then further confirmed with isotope labelling. Aiming to optimize our catalysts in the following years, we have founded Atmonia ltd. for developing the process and business model.

536265 Electrochemical Synthesis of Ammonia Using Nitrogen and Water in Alkaline Electrolytes Under Ambient Conditions

Abstract:

Sustainable synthesis of Ammonia (NH_3) is gaining great attention not only for its use as an alternative renewable energy fuel but also to substitute production of distributed fertilizers through the conventional Haber Bosh process. The conventional Haber-Bosh process to produce NH_3 uses fossil fuels in deriving hydrogen from steam reforming of natural gas, is energy intensive and also leads to significant CO_2 emission. Alternatively, electrochemical synthesis of ammonia (ESA) through the nitrogen reduction reaction (NRR) in alkaline medium saves the use of hydrogen as a reactant as the aqueous electrolyte forms the source of proton. However, the standard reduction potential of nitrogen and hydrogen fall in the same domain. Thus, hydrogen evolution reaction is so dominant at the applied potential that selectivity of nitrogen reduction is a major challenge in the budding field. The rate-determining step of the NRR is predicted to be the adsorption and bond activation of the N_2 molecule. Herein, we report a comparative study of nitrogen adsorption on basic sites of different metal-organic framework. For certain MOFs, the central metal can be a basic site and for some others, the interplay between the organic ligand and the metal center can create basic sites. Despite the basic sites, it might not be easy to selectively reduce nitrogen in presence of protons in the electrolyte at the metal centers. We observed that MOF derived nitrogen-doped metal free nanoporous carbon electrocatalyst has a Faradaic efficiency (FE) greater than 10 % at -0.3 V *vs* RHE under ambient conditions for the NRR. ZIF8 derived nanoporous carbon exhibits a remarkable production rate of NH_3 up to 3.4×10^{-6} mol cm^{-2} h^{-1} using aqueous 0.1 M KOH electrolyte. The performance has been compared with other N doped carbon derived from commercial polyaniline, nitrogen free KJ black and N doped CNT. The stability of the nitrogen-doped electrocatalyst was demonstrated during an 18-hour continuous test with constant production rates. In Ar a similar test was done to confirm that the ammonia is not derived from the catalytic decomposition. First principles calculations were used to elucidate the possible active sites, suggesting that the moiety consisting of three pyridinic N atoms adjacent with a nitrogen vacancy embedded in a carbon layer is able to strongly adsorb N_2 and further realize N°N bond dissociation for the subsequent protonation process. We have further introduced precious metals like Au, Ru and Pt into different MOFs and investigated their activity compare to precious metal metal free MOF. Au is found to be much more active compared to other precious metals. This work provides a new insight into the rational design and synthesis of nitrogen-doped and defect-rich carbon NRR catalysts for NH_3 synthesis at ambient conditions.

545770 Analysis of Influence of Operating Pressure on Dynamic Behavior of Ammonia Production over Ruthenium Catalyst under High Pressure Condition

Abstract:

Process technologies on energy conversion of renewable electricity into hydrogen energy carrier are significant to deploy long-term storage and long-distance transport of much more renewable inside and outside Japan. Ammonia is a potential hydrogen carrier that contains 17.6 wt% of hydrogen. Moreover, as an energy carrier, ammonia is thought to be a clean fuel as only water and nitrogen are produced on direct combustion.

Many researchers and engineers consider that ammonia plants using hydrogen produced by solar electricity or wind electricity will be much smaller than those currently used [1]. There is an issue of low pressure condition for feed of raw material gas, since hydrogen and nitrogen are produced by water electrolysis and pressure swing adsorption (or cryogenic air separation) respectively. Ammonia synthesis under the conventional pressure conditions needs increase in power for compression of the feed gas. Moreover, low temperature operation has advantage in getting higher ammonia concentration of equilibrium limitation. In order to promote the reaction under the lower pressure conditions, lower reaction temperature condition is desirable due to a limitation of equilibrium.

In Japan, the Japan Science and Technology Agency (JST) supports the research and development of catalyst with high catalytic activities in the low-temperature region, and our research group investigates performance of developed ruthenium catalysts by small-scale ammonia plant that is built in Fukushima Renewable Energy Institute, AIST (FREA). Ruthenium (Ru) catalysts have received particular attention owing higher activity for ammonia synthesis, since the pioneering study of Aika et al. [2] Great efforts have been paid to develop supported Ru catalysts using various metal supports that are Al_2O_3, MgO, Cs-MgO and so on. However, there is a small number of studies for influence of time-variation of flow rate of hydrogen and operating pressure to kinetics of ammonia synthesis by using the Ru catalysts supported on metal oxides, from the viewpoint of design of the small-scale ammonia production systems by using renewable energy.

In the NH_3 fuel conference of last year, we have reported influence of flow ratio of hydrogen to nitrogen (γ) to dynamic behavior of ammonia production by using the developed Ru/CeO_2 under low pressure conditions (< 2.5 MPa). The highest value of space time yield (STY) for ammonia production was seen when temperature (T) is 375 °C and γ =1.5. Then it was seen that the STY increased linearly with the operating pressure when T = 400 °C. We expected that increase in operating pressure could enhance reactivity of Ru/CeO_2 in a case when T = 400 °C, however dynamic behavior of ammonia production over the Ru/CeO_2 under high pressure conditions (> 2.5 MPa) was not made clear. In the present study, we analyze influence of operating pressure to dynamic behavior of ammonia production over the Ru catalysts. Characteristics of reaction behavior on Ru/CeO2 will be discussed by using different types of macro kinetic models (Langmuir-Hinshelwood (LHHW) equation models).

Laboratory kinetic experiments for the developed Ru catalyst was carried out by using a quartz tubular reactor (φ14 mm x 370 mm). The catalyst of about 2 milliliters was packed in the middle. As the preliminary step to activate the catalyst, hydrogen and nitrogen was flowed in a stoichiometry ratio for 30 minutes under condition of 600 ºC and ambient pressure. In the present study, ammonia synthesis was carried out by changing flow ratio of hydrogen to nitrogen (γ) in a range of 0.25 – 4.0, temperature in a range of 325 – 425 °C, gage pressure (P) in a range of 0.0 – 10 MPa. The produced ammonia was analyzed by the decrease of electron conductivity in the sulfuric acid solution.

First, data of STY for ammonia production by using Ru-Cs/MgO under low pressure conditions (< 2.5 MPa) were analyzed by the Temkin equation model that was well-known, which could be derived under the assumption that the reaction rate-determining step (RDS) was dissociative adsorption of nitrogen. In the present paper, we apply plots for relationships between the factor of adsorption term (denominator of LHHW equation) and the ratio of driving force (numerator of LHHW equation) to rate of ammonia production [3], in order to investigate applicability of numerical formula of kinetic model that is changed by assumption for the RDS and competitive adsorption of intermediates on the active site. When temperature was 400 ºC, plots for various flow ratio γ showed tendency of linear increase with value of (pA / pH1.5) that was changed by operating pressure, which indicated applicability of the Temkin equation model. However, in a case when T = 325 ºC, a tendency of plots that was different from 400 ºC was seen. Since value of (pA / pH1.5) decrease with increase in operating pressure, it was considered that a massive surface coverage with adsorbed hydrogen inhibiting the catalytic activity was induced by increasing pressure. Then, it was considered that the interaction between hydrogen and Ru and or Cs on the surface of the catalyst become stronger at 325 °C.

Next, data of STY for ammonia production by using Ru/CeO2 under high pressure conditions (<10 MPa) were analyzed by using the power law expression (r = k pN2l pH2m pNH3n). In this kinetic analysis, the exponent of partial pressure for hydrogen (m) was positive. Then, the exponent of partial pressure for nitrogen (l) was 0.37 smaller than 1. According to the previous paper [4] for kinetic analysis for Ru-Cs/MgO, the l's value of 1 can be derived from the Temkin model. Since it is supposed that formula of the LHHW equation model for Ru/CeO2 may be different from the Temkin equation model, we investigate application of other kinetic model based on the assumption of competitive adsorption of hydrogen radical and nitrogen hydride radicals on the active site, by using the above-mentioned analytical plots for relationships between the factor of adsorption term and the ratio of driving force to rate of ammonia production.

When data of STY for a case when T = 400 °C and γ = 3 were analyzed by the analytical formula based on the Temkin model, tendency of plots with negative slope in a region of high pressure (P > 4 MPa) showed that conventional assumption of competitive adsorption of hydrogen radical and nitrogen radical on the active site was not applicable to interpretation of reaction behavior on Ru/CeO2 catalyst. On the other hand, LHHW equation model based on the assumption of competitive adsorption of hydrogen radical and NH2 radical on the active site showed useful for the kinetic analysis, due to tendency of linear increase with value of (pA / pH0.5) that was changed by operating pressure. Therefore, it was considered that competitive adsorption of hydrogen radical and nitrogen hydride radicals was a key factor for retardation of progress of ammonia production by using Ru/CeO2. In future work, influence of adsorption nitrogen hydride radicals under high pressure condition will be analyzed by data of STY that were collected by changing the flow ratio of hydrogen to nitrogen.
This work was supported by Council for Science, Technology and Innovation (CSTI), Cross-ministerial Strategic Innovation Promotion Program (SIP), "Energy carriers" (funding agency: Japan Science and Technology Agency, JST).

References
[1] M. Malmali et al., Ind. Eng. Chem. Res., 55 (2016) 8922-8932.
[2] K. Aika et al.; J. Catal., 92 (1985) 296-304.
[3] H. Kubota and M. Shido; Kagakogaku, 23 (1959) 242-249.
[4] K. Aika; Catal. Today, 286 (2017) 14-20.

522157 Test Results of the Ammonia Mixed Combustion at Mizushima Power Station Unit No.2 and Related Patent Applications

Abstract:

Test results of the ammonia mixed Combustion at Mizushima Power Station Unit No.2 and related patent applications. At the Mizushima Power Station Unit No.2(Coal-fired, Location: Kurashiki, Okayama Prefecture, rated output: 156,000 kW), the Chugoku Electric Power Company conducted the ammonia mixed-combustion test from July 3 to 9 in order to reduce the environmental burden of coal-fired power stations. We compile the test results and report it to Japan Science and Technology Agency(JST), and we are pleased to inform you today that we applied for patents on the findings obtained in this examination. This test is part of the "ammonia direct burning" as a commissioned research topic on "energy carriers" out of the "Strategic Innovation Promotion Program (SIP)" of the Cabinet Office's Science and Technology and Innovation Council, commissioned from JST. In the test, it was confirmed that ammonia used as fuel could be burned completely in the boiler and contributed to power generation. In addition, it was not confirmed the environmental problems caused by nitrogen oxides. Initially, the co-firing rate was set at about 0.6% by operating the generator output at 155,000 kW and maximally supplying ammonia from the existing ammonia facility. However, due to the weather, by lowering the generator output to 120,000 kW, it was possible to burn the mixture to about 0.8% (equivalent to 1,000 kW) as a result. Since we have confirmed new knowledge that the concentration of nitrogen oxide tends to decrease under certain conditions in the combustion method tested this time, we have applied for a patent. Through this examination, we confirmed that ammonia can contribute to power generation as a fuel, without greatly affecting the environment, with regard to co-firing of ammonia to coal-fired power plants. Therefore, it is possible to further increase of the ammonia co-firing rate for coal-fired power plant in the future. Chugoku Electric Power Co. and JST will continue to contribute to reducing the environmental impact of thermal power stations and energy efficiency. * 1: Strategic Innovation Promotion Program (SIP): A national project to realize science and technology innovation by The Cabinet Office's Integrated Science and Technology ï½¥ Innovation Council having the function of the command tower, with the management beyond the framework of the Ministry of Council and the old field. * 2: Energy carrier: A substance used for energy transportation / storage. For the hydrogen energy, there are liquefied hydrogen, organic hydride (methyl -cyclohexane), ammonia and the like, and researches are progressing on each.

521784 Performance of Ammonia - Natural Gas Co-Fired Gas Turbine for Power Generation

Abstract:

Performance of Ammonia - Natural Gas Co-fired Gas Turbine for Power Generation

Ammonia is paid special attention as renewable energy carrier [1-3], because it offers advantages in generation, transportation and utilization. Harbor-Bosch method is already established as ammonia generation method; large amount of ammonia is already used as fertilizer and chemical raw material. Ammonia can be liquefied at room temperature. Its transport and storage system are already established. Ammonia is cheaper to transport than hydrogen. Ammonia can be used as carbon-free fuel in internal combustion engines as alternative to conventional hydrocarbon fuels. However, it has different combustion characteristics. For example, the nitrogen atom contained in the ammonia molecule, causes high NOx emission [3]. It might be difficult to achieve stable combustion when fueling ammonia to an internal combustion engine due to low laminar burning velocity. It also might cause large emission of unburnt components [4]. Many efforts have been devoted to overcome these shortcomings [5-7]. Especially, Iki et al. demonstrated the successful power generation by an ammonia-fueled 50 kWe micro gas-turbine for the first time [8, 9]. Performance of the gas turbine revealed a combustion efficiency of 89-96%, when only unburned ammonia is accounted for; NOx emission was above 1000 ppm@16%O_2. In the present study, demonstration of ammonia – natural gas co-firing in a larger gas turbine is examined. Strategies of low NOx combustion reported in past studies [10-12], are also adopted in engine testing.

For the demonstration, IM270, a simple cycle gas turbine manufactured by IHI Corporation [13], was used. The test system consists of Selective Catalytic Reduction (SCR) unit, natural-gas compressor and ammonia supply unit. The fuel supply unit first pressurizes ammonia to 2 MPaG and then gasifies it in a steam vaporizer, before releasing it to the combustor. In engine testing, the gas turbine is first started and then power is increased up to 2 MWe power generation output firing natural gas, before ammonia is supplied to the combustor. Ammonia supply to the engine is measured in terms of the heat input ratio of ammonia to total fuel. This ratio is called "ammonia mixing ratio" in this study. Operation of the gas turbine engine turned out to be stable in the whole range of ammonia mixing ratios from zero to 20%.

Figures 1 and 2 show the performance of the gas turbine engine. It is seen that CO_2 concentration at turbine outlet monotonously decreases as ammonia mixing ratio is increased. By increasing the ammonia mixing ratio from zero to 20% in LHV, CO_2 concentration is reduced from 3.1% to 2.5%, i.e. by 0.6%. It shows that ammonia supplied to the gas turbine combustor is converted to power, so that the amount of natural gas required for 2 MWe power generation is decreased. It is also found that CO and unburnt ammonia concentration at the turbine outlet are lower than the detection limit of the measuring equipment. As ammonia mixing ratio is increased, NOx concentration at the turbine outlet first drastically increases up to a mixing ratio of 5%, then, remains constant until it reaches 20%. NOx emission is 287 ppm@16%O_2 at ammonia mixing ratio of 20%, which is much higher compared to typical natural-gas fired gas turbines. However, it is shown that NOx emission can be reduced below 6 ppm@16%O_2 by the SCR unit. Combustion efficiency is above 99.8% for all test conditions. It is to be noted that in the evaluation of combustion efficiency, loss of effective calorific value due to NOx emission is accounted for in addition to CO and unburnt ammonia emission. It shows that amount of NOx emission mentioned above does not have a strong impact on combustion efficiency. It is also found that generator-end efficiency takes smallest value at ammonia mixing ratio of 5%, then monotonically increases with increasing ammonia mixing ratio. The decrease of generator-end efficiency at ammonia mixing ratio of 5% is due to rapid increase of NOx emission. Although the impact of NOx emission on combustion efficiency is small, rapid increase of NOx emission, which means loss of effective calorific value, causes a loss of 0.6% in generator-end efficiency. On the other hand, generator-end efficiency is increased when ammonia mixing ratio is increased above 5%. This result is believed to be caused by the increase of gas volume, when fuel is changed from natural gas to ammonia. As ammonia mixing ratio is increased, total gas volume passing through the turbine increases, which leads to increased workloads in the turbine. Then, total fuel heat input required to maintain 2 MWe output is expected to become smaller. For more details, energy balances of the gas turbine engine cycle should be investigated.

First shot test results show that ammonia can be used as gas turbine fuel in a 2 MWe gas turbine engine. However, reduction of NOx emission is important for reducing running costs, which are increased by the ammonia used in the SCR unit. Therefore, further research is required to develop a low NOx combustor.

This work was supported by Council for Science, Technology and Innovation (CSTI), Cross-ministerial Strategic Innovation Promotion Program (SIP), "Energy Carriers" (Funding agency: JST).

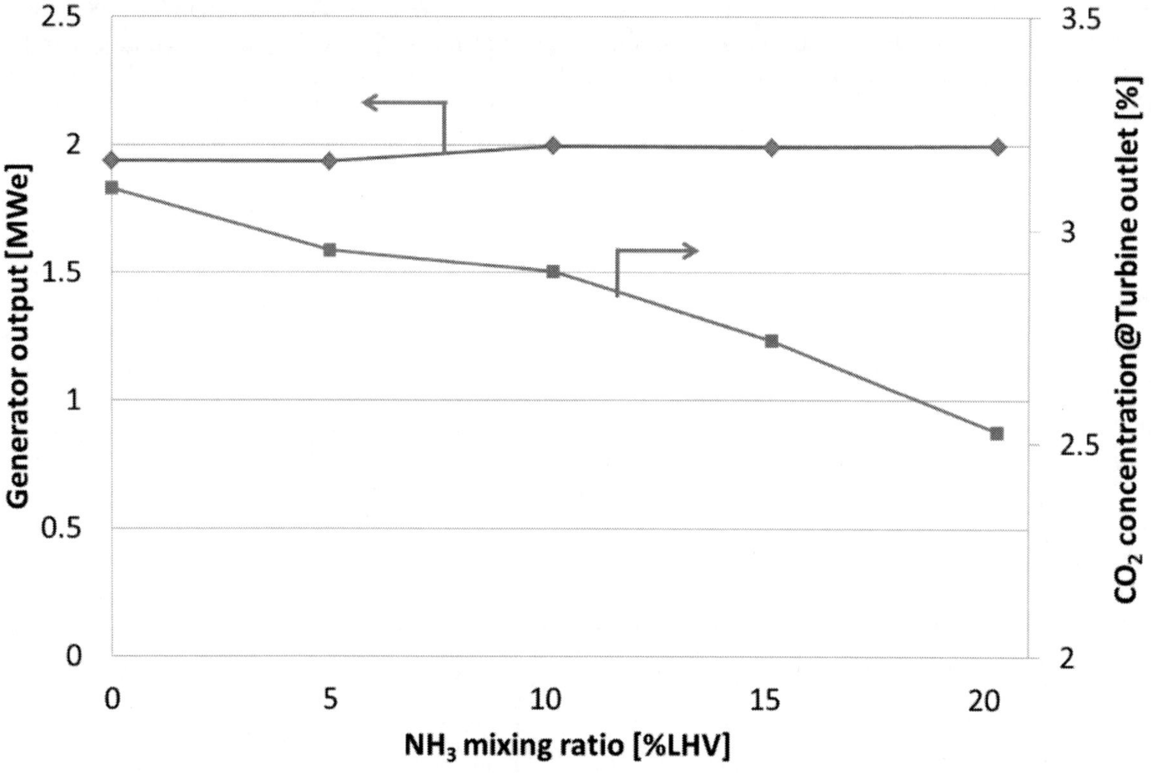

Fig. 1 Relationship among ammonia mixing ratio, generator output and CO_2 emission at turbine outlet.

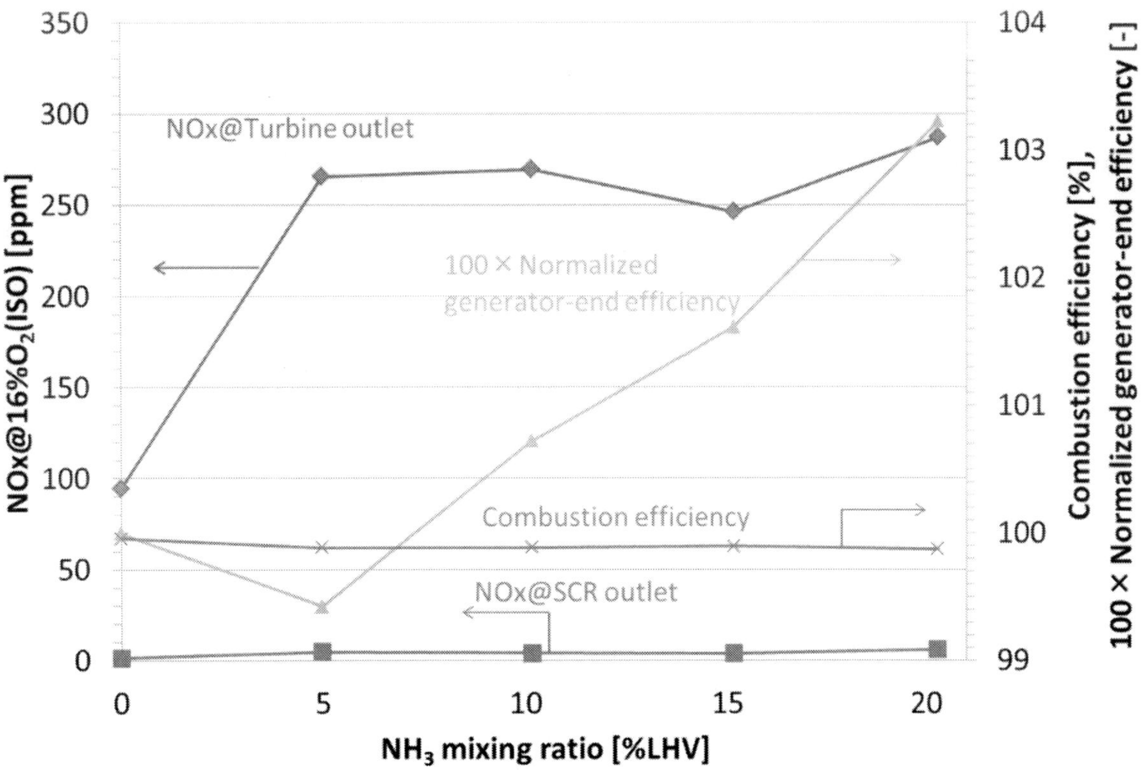

Fig. 2 Relationship among ammonia mixing ratio, NOx emission and normalized generator-end efficiency.

Reference

[1] C. Zamfirescu, I. Dincer, J. of Power Sources 185 (2008) 459-465.

[2] E. A. Gilmore, A. Blohm, S. Sinasabaugh, Renewable Energy 71 (2014) 361-367.

[3] P.Trop, D. Goricanec, Energy 108 (2016) 155-161.

[4] A. Hayakawa, T. Goto, R. Mimoto, T. Kudo, H. Kobayashi, Mech. Eng. J., 2(1), (2015), 14-00402.

[5] A. Hayakawa, T. Goto, R. Mimoto, Y. Arakawa, T. Kudo, H. Kobayashi, Fuel, 159, (2015), 98-106.

[6] A. Hayakawa, Y. Arakawa, R. Mimoto, K.D. Kunkuma, A. Somarathne, T. Kudo, H. Kobayashi, J. of HYDROGEN ENERGY 42 (2017) 14010-14018.

[7] A. Valera-Medina, S. Morris, J. Runyon, D. G. Pugh, R. Marsh, P. Beasley, T. Hughes, Energy Procedia 75 (2015) 118-123.

[8] A. Valera-Medina, R. Marsh, J. Runyon, D. Pugh, P. Beasley, T. Hughes, P. Bowen, Applied Energy 185 (2017) 1362-1371.

[9] N. Iki , O. Kurata , T. Matsunuma , T. Inoue , M. Suzuki , T. Tsujimura , H. Furutani , Proc. of ASME Turbo Expo 2015 GT2015-43689, Montreal.

[10] O. Kurata, N. Iki, T. Matsunuma, T. Inoue, T. Tsujimura, H. Furutani, H. Kobayashi, A. Hayakawa, Proc. Combust. Inst. 36 (2017) 3351-3359.

[11] S. Ito, S. Kato, T. Saito, T. Fujimori, H. Kobayashi, Ammonia fuel conference 2015, (2015)

[12] S. Ito, S. Kato, T. Saito, T. Fujimori, H. Kobayashi, Ammonia fuel conference 2016, (2016)

[13] S. Onishi, S. Ito, M. Uchida, S. Kato, T. Saito, T. Fujimori, H. Kobayashi, 2017 AIChE Annual Meeting, (2017)

[14] IHI Corporation home page, https://www.ihi.co.jp/powersystems/en/lineup/IM270/index.html

540654 Rapid Ramp NH_3 Prototype Reactor Update

Abstract:

Starfire Energy has built and operated a low pressure, fast-ramping prototype reactor using its Rapid Ramp NH_3 process. It has synthesized, captured, and liquefied NH_3 with all system pressures staying below 12.5 bar. The prototype reactorâ€™s performance will be discussed.

540499 Realisation of Large-Scale Green Ammonia Plants

Abstract:

The global ammonia production is nowadays mostly based on fossil energy carriers (natural gas, coal, naphtha, etc.). It consumes approximately 1.4% fossil energy carriers and releases more than 1.4% of global CO2 emissions.

In order to continue the global transition from the fossil fuel and nuclear energy age to the renewable energy age, ammonia could play a key role. Beside the continued utilization for fertilizer industry, ammonia could become an energy and/or hydrogen carrier as well.

thyssenkrupp Industrial Solutions (tkIS) developed a concept to establish Green Ammonia Plants as an alternative to conventional world-scale ammonia plants. As industry leader in electrolysis (AWE technology) and ammonia business (uhde® ammonia synthesis), tkIS combines the knowledge in both technologies to offer electricity-based ammonia plants in the near future.

540695 Demonstration of CO_2-Free Ammonia Synthesis Using Renewable Energy-Generated Hydrogen

Abstract:

In Japan, the government funding project SIP, Strategic Innovation Promotion Program, supports the research, development and demonstration of "Energy Carriers". The concept of the "Energy Carriers" value chain is to produce hydrogen energy carriers overseas from fossil resources using CCS or renewable energy, and transport it to Japan for utilization as clean energy. The purpose of the program is to help realize a low-carbon society in Japan by using hydrogen. Among energy carriers, ammonia is the one of the most promising carriers, because of the ease of transportation as a liquid, higher hydrogen density, and proven technologies for commercial and industrial scale, not only for production, storage, and transportation, but also its utilization in chemical plants and DeNOx units for electric power plants.

Under the theme of "Development of ammonia synthesis from CO_2-free hydrogen" of SIP "Energy Carriers", JGC is developing the advanced ammonia synthesis process using renewable energy, such as Photovoltaic and Wind Turbine Power Generation, to be able to produce "Green" ammonia, aiming to contribute to a low-carbon society. Also, utilizing the catalysts developed by the National Institute of Advanced Industrial Science and Technology (hereinafter "AIST"), National Institute of Technology Numazu College, and JGC C&C, AIST and JGC designed and constructed an ammonia synthesis demonstration plant in FREA, the Fukushima Renewable Energy Institute, AIST by the end of fiscal year 2017. From April 2018, the plant started operation to evaluate the performance of the developed catalysts and acquire the engineering data for scaling up in the future.

In this paper, we would like to explain the details of the ammonia synthesis demonstration plant in FREA, such as process flow, plant operation conditions, its capacity, and the status of plant operation.

One of the key features of the developed catalyst is its high performance at lower temperature and lower pressure than the conventional Haber-Bosch process. The lower operation pressure contributes to lower CAPEX and OPEX for ammonia production using renewable energy.

Also, we will explain our future plan for the CO_2-free ammonia value chain demonstration, which is from renewable energy power generation, hydrogen production, and ammonia synthesis to ammonia direct power generation at the same site location, since FREA already owns Photovoltaic, Wind Turbine, Water Electrolysis, and Ammonia direct combustion Micro Gas Turbine research facilities.

This demonstration project is supported by the Council for Science, Technology and Innovation (CSTI), Cross-Ministry Strategic Innovation Promotion Program (SIP), "Energy Carriers" (Funding Agency: Japan Science and Technology Agency).

530473 Design Optimization of an Ammonia-Based Distributed Sustainable Agricultural Energy System

Abstract:

Small-scale, distributed production of ammonia better enables the use of renewable energy for its synthesis than the current paradigm of large-scale, centralized production. Pursuant to this idea, a small-scale Haber-Bosch process has been installed at the West Central Research and Outreach Center (WCROC) in Morris, MN [1] and there is ongoing work on an absorbent-enhanced process at the University of Minnesota [2], [3]. Using renewables to make ammonia would greatly improve the sustainability of fertilizer production, which currently accounts for 1% of total global energy consumption [4]. The promise of renewable-powered, distributed ammonia production for sustainability is in fact not limited to fertilizer, because ammonia also has potential as an energy-dense, carbon-neutral fuel. For example, using ammonia produced from renewable energy for nitrogen fertilizer, grain drying fuel and tractor fuel at the WCROC farm would reduce more than 90% of the fossil energy footprint associated with corn production [5].

In this light, we envision a distributed sustainable agricultural (farm) energy system (DSAE) fundamentally based on the idea of ammonia as not only a fertilizer, but also a fuel and a method of energy storage. Specifically, this system will use only renewable energy to produce ammonia for use as fertilizer and agricultural fuel (for cropping equipment and grain drying) at the scale of a single farm or an agricultural cooperative. It will also use renewables to meet local power and heat demands in a manner synergistic to distributed ammonia production; the difference in power and heat (hourly) and ammonia (monthly or biannually) demand time scales gives rise to opportunities for temporally flexible ammonia production and locally controllable power generation using ammonia. Heat integration will also be possible due to the exothermic nature of ammonia synthesis.

Based on the above, we consider a DSAE superstructure that consists of PV arrays, wind turbines, a battery bank, electrolysis (for hydrogen production), pressure-swing adsorption (for nitrogen production) and modular ammonia synthesis units, hydrogen, nitrogen and ammonia storage vessels, hydrogen- or ammonia-fed fuel cells, ammonia-fed microturbines and electric boilers. Additionally, excess power can be sold to the utility to generate revenue, but this sale is constrained to prevent sharp inter-hour changes which can adversely affect grid operation.

For this superstructure, we develop a framework for design optimization. The 20 year net present cost (NPC) of this system is minimized by determining the optimal selection and size of units. The design of this system is inextricably coupled with its scheduling due to the time-varying nature of renewable generation as well as power and heat demands. Therefore, we consider one year of operation during which set points of installed units are allowed to vary hourly. For the sake of computational tractability, 12 representative weeks (2016 hours), one for each month, are used in the place of a full year-long scheduling horizon (8760 hours). The operating cost over 20 years is determined by first assuming identical operation in each month from each representative week and then assuming identical operation in each year from the scheduling year (scaled using a discount rate). This combined design-scheduling problem is formulated as a mixed-integer linear program (MILP) through the use of piecewise linear capital cost correlations and linear production-consumption relationships for unit operation.

We apply this framework to a case study inspired by the WCROC farm and the nearby UMM Morris Campus. The farm consists of 290 acres of corn and 116 acres of soy. The UMM Campus has hourly average power and heat loads of 985 kWh and 3480 kWh, respectively. UMM strives to be a net-zero energy campus by 2020 and to this end, two 1.65 MW wind turbines, a 20 kW solar PV system and biomass gasification district heating and cooling system with a 300 kW back pressure steam turbine have been installed so far. For this case study, we assume that the DSAE system can utilize the already-installed wind turbines and PV array while also having the ability to install additional renewable generation capacity. We further assume that renewable ammonia is required as fertilizer and as fuel for cropping operations and grain drying for all WCROC crops and that the DSAE system is responsible for serving the power and heat loads not met by biomass gasification and the associated back pressure steam turbine. Hourly power and heat demands as well as the ammonia demand profile are based on historical data from the Morris-WCROC site. Renewable generation profiles are synthesized using relevant weather data (e.g. insolation, ambient temperature, wind speed) for a typical meteorological year in Morris, MN. Based on the results of this case study, we perform sensitivity analysis to determine the unit performance improvements or capital cost reductions that will be most effective in reducing the NPC of the DSAE system.

[1] M. Reese, C. Marquart, M. Malmali, K. Wagner, E. Buchanan, A. McCormick and E. L. Cussler, "Performance of a Small-Scale Haber Process," *Industrial & Engineering Chemistry Research,* no. 55, pp. 3742-3750, 2016.

[2] M. Malmali, Y. Wei, A. McCormick and E. L. Cussler, "Ammonia Synthesis at Reduced Pressure via Reactive Separation," *Industrial & Engineering Chemistry Research,* vol. 55, no. 33, pp. 8922-8932, 2016.

[3] K. Wagner, M. Malmali, C. Smith, A. McCormick, E. Cussler, M. Zhu and N. C. Seaton, "Column Absorption for Reproducible Cyclic Separation in Small Scale Ammonia Synthesis," *AIChE Journal,* vol. 63, no. 7, pp. 3058-3068, 2017.

[4] B. Swaminathan and K. Sukalac, "Technology transfer and mitigation of climate change: The fertilizer industry perspective," in *InIPCC Expert Meeting on Industrial Technology Development, Transfer and Diffusion*, Tokyo, 2004.

[5] J. Tallaksen, *Life Cycle Assesment of Corn Grain at WCROC,* Morris, MN: Morris SunTribune, 2016.

Evaluation of the cement clinker fired in the combustion furnace of heavy-oil and NH_3

Hiroki Kujiraoka[1], Tatsurou Izumi[1], Yuya Yoshizuru[1], Takeshi Suemasu[1], Makoto Ueda[1], Toyoaki Niki[1], Takayasu Itou[1], Masayuki Nishio[1], Ryuichi Murai[2] and Fumiteru Akamatsu[2],
(1)UBE Industries, LTD., UBE, Japan,
(2)Mechanical Engineering, Osaka University, Osaka, Japan

Introduction

The large-scale use of renewable energy which doesn't generate CO_2 is expected to solve the problem such as depletion of fossil fuel and global warming. A secondary energy with low energy load, high energy density and no CO_2 generation is necessary to store and transport renewable energy. Hydrogen is a secondary energy of low environmental load, but a high pressure is necessary to achieve high density. So, many people have researched and developed hydrogen absorbing alloys, organic hydrides and so on as hydrogen carriers for storing hydrogen lightly and compactly. Among them, ammonia (NH_3) has advantage in consistent developed technology of production, transportation and storage, high hydrogen storage, and possible direct combustion. Therefore, NH_3 is expected as a hydrogen energy carrier.

Primary energy consumption in the industrial furnace field accounts for 17% of the primary energy consumption in Japan [1]. Introducing NH_3 as a primary energy source in the industrial furnace field is a technical development task with great potential in CO_2 reduction. Based on the average coal usage for domestic cement production in 5 years from FY 2011 [2], when 1% coal is replaced with NH_3 on the basis of calorific value, CO_2 reduction amount is estimated to be 160,000 tons/year, which corresponds to 1% CO_2 reduction on a weight basis.

Assuming application of NH_3 mixed combustion in a cement kiln, a radiative heat transfer decreases due to a decrease in CO_2 concentration as compared with coal or heavy-oil combustion. So the temperature of the raw materials of the cement clinker as the heated object is also expected to decrease. In a cement manufacturing process, there are examples of using NH_3 spray and NH_3 nitrogen waste for de-nitration of the exhaust gas, but there is no example using NH_3 as a thermal energy source.

Therefore, we have researched [A] Influence of atmosphere conditions on the cement clinker formation rate [B] NH_3 mixed combustion for example heavy-oil – NH_3 in the test furnace [C] Evaluation of the cement clinker fired under NH_3 mixed combustion in the test furnace [D] Process simulation of cement clinker production under NH_3 mixed combustion, respectively.

In this report, we would like to explain two basic studies ([A] and [C]) for developing the technology to apply NH_3 mixed combustion in a cement kiln. First, we burned the clinker in the atmosphere-controlled electric furnace to investigate the influence of gas composition on the clinker formation reaction rate. Second, we evaluated the quality and characteristics of the clinker actually fired under NH_3 mixed combustion in the test furnace.

Experimental

Materials

Raw materials were mainly used as same as those used in the actual plant, and reagent partially was used. Calcium carbonate ($CaCO_3$) and silica stone powder were obtained from own company's plant use. Fry ash was obtained from JIS II fry ash. Fe_2O_3 (Purity >95%), $CaSO_4 \cdot 2H_2O$ (Purity >98%), Na_2CO_3 (Purity >99.5%), K_2CO_3 (Purity >99.5%) was purchased from FUJIFILM Wako Pure Chemical Corporation. Table 1 shows the composition of raw materials expect reagent.

Raw materials were mixed in the composition as same as ordinary Portland cement clinker. Those and pure water were mixed, and poured into the mold. Those were dried at 105 °C for half a day, after then cement clinker pellets were obtained which was about 1 square centimeter. Table 2 shows the mixing ratio of raw materials for preparing cement clinker pellets.

Table 1. Composition of raw materials expect reagent for preparing cement clinker pellets

	CaO	SiO_2	Al_2O_3	Fe_2O_3
	[wt. %]			
$CaCO_3$	55.45	0.06	0.02	0.02
Fry Ash 1	2.8	61.0	24.2	5.17
Fry Ash 2	3.3	64.7	20.8	5.04
Silica stone powder	0.36	94.16	1.96	1.63

Table 2. Mixing ratio of raw materials for cement clinker pellets

	Experiment 1	Experiment 2
	[kg]	
$CaCO_3$	11.7	11.7
Fry Ash 1	2.34	0.00
Fry Ash 2	0.00	2.76
Silica stone powder	0.845	0.466
Fe_2O_3	0.198	0.185
$CaSO_4 \cdot 2H_2O$	0.172	0.172
Na_2CO_3	0.051	0.051
K_2CO_3	0.117	0.117

Burning clinker in atmosphere-controlled electric furnace (Experiment 1)

The procedure of atmosphere control in electric furnace (S6V-1515D, Motoyama Co. Ltd.) was as follows. CO_2, Air, and N_2 gas were supplied from the gas cylinder, H_2O was only supplied from the superheated steam generator (SH-ROBO, Nippon Denki Co. Ltd.). Assuming fully combustion of NH_3 mixed energy, the gas composition flowed into the electric furnace was determined, and was shown in Table 3.

Table 3. Gas composition assuming NH_3 mixed combustion in experiment 1

	Mixed NH_3 (cal. base %)	Gas Conditions (vol. %)			
		CO_2	H_2O	N_2	O_2
Test I	0	16.4	6.3	73.4	3.9
Test II	30	10.7	14.6	71.1	3.6

The raw mixture was calcined at 1000 °C for 30 min, at the same moment, the gas was flown into the furnace according to prescribed conditions (Table 3). After then those were heated up at 25 °C/min and burned at 1350, 1450, or 1550 °C for 0, 10, or 30 min respectively. Figure 1 shows the temperature profile. It was confirmed beforehand that the gas flow rate had no effect on the furnace temperature control.

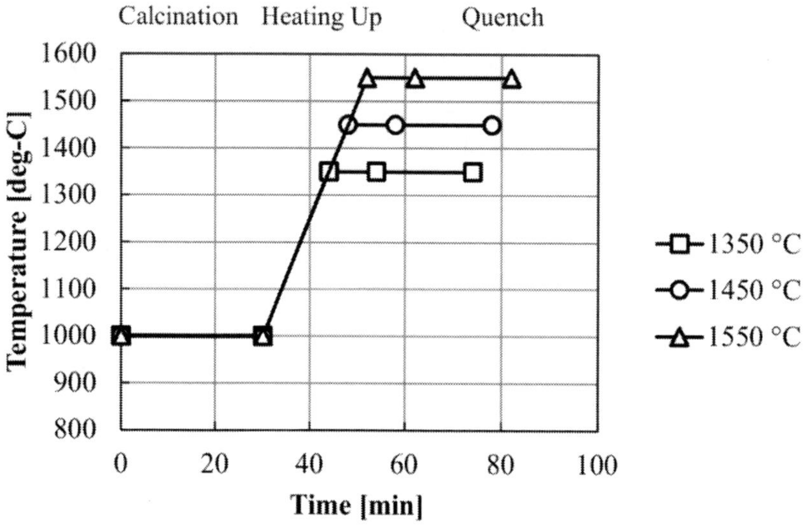

Figure 1. Temperature profile in experiment 1; Open square, 1350 °C, Open circle, 1450 °C, Open triangle, 1550 °C.

Firing clinker under NH_3 mixed combustion in 10kW furnace (Experiment 2)

The inner of 10 kW furnace at Osaka University was 1200 mm long, 300 wide, and 300 mm high, but it was extended by 300 mm long to take out the clinker sample. The inside temperature of the furnace was controlled by the thermocouple inserted from the top wall surface, and NOx, CO, CO_2, O_2 concentration and temperature in the exhaust gas was measured. The burner had a quadruple tube structure, and a tip for spraying heavy-oil in the innermost tube. The triple pipe section on the outer periphery of the pipe could independently supply the air, we defined the inside as 0th order (0-Air), the midst as 1st (1-Air) and the outside as 2nd (2-Air), respectively.

The procedure of the experiment was as follows. First, 2 kg clinker pellets on the two of platinum tray were injected into the furnace. Second, the inner furnace was preheated using a pilot burner by city gas until reaching 600 °C, after then heavy oil was started to supply at 32.3 mL/min (equivalent to 20 kW).

Under only heavy-oil firing condition, it was heated up and kept for 30 minutes after temperature of the thermocouple in contact with platinum tray reached 1400 °C. Under NH_3 mixed firing condition, after the burner-side furnace temperature reached 1000 °C, the supply of heavy-oil was reduced and NH_3 was started to supply. After the mixing rate of NH_3 was increased up to 30% on the basis of calorific value, it was also heated up at 1400 °C and then kept for 30 minutes under output condition of 20 kW.

After stopping to hold temperature in all conditions, heavy-oil and NH_3 was stopped to supply, and room temperature air of 120 NL/min was flowed into the furnace to quench the sample. After the temperature inside the furnace reached 500 °C or lower, the air was stopped to supply and the sample was left to cool overnight, taken out.

Chemical Analysis of Clinker

The chemical composition of clinker was analyzed according to JIS R 5204 [5], and the free lime (f-CaO) was analyzed according to JSACI-01 [6]. The mineral composition was determined by XRD/Rietveld analysis.

Calculation of clinker reaction rate

Cement clinker reactions are largely classified into Blite formation ($2CaO + SiO_2 \rightarrow 2CaO{\cdot}SiO_2$ (C2S)) and Alite formation ($CaO + C2S \rightarrow 3CaO{\cdot}SiO_2$ (C3S)). These solid phase reaction rates are determined by the diffusion rate of the reactant and represented by Jander equation [3].

$$(1-\sqrt[3]{1-\alpha})^2 = Kt \tag{1}$$

α is the CaO reaction rate (-), t is the reaction time (min) and K is the reaction rate constant represented by the Arrhenius equation.

$$K = A\exp\left(\frac{-E}{RT}\right) \tag{2}$$

Here, A is the frequency factor (1/min), E is the activation energy (J/mol), R is the gas constant (8.314 J/mol·K), and T is the reaction temperature (K).

In this case for comparing simply the reaction rate constant, A was only calculated and E was defined as 29,920 J/mol calculated previously for ordinary Portland cement clinker.

Physical Testing of Cement

For physical property experiments, a predetermined amount of exhausted and di-hydrated gypsum was added to calcined clinker, and then those mixtures were milled by the ball to prepare cement. Physical properties were tested according to JIS R 5201 [7], and the mortar compressive strength was measured at 3 days, 7 days and 28 days of age.

Results and Discussion

Influence of Gas Conditions on Clinker Reaction Rate

The clinker became darker as the temperature increased, looking at the temperature. On the other hand, there is no significant difference in the appearance, looking at the atmosphere.

Next, Table 4 shows the composition analysis results of the burned clinker. Regardless of the atmosphere condition, formed alite (C3S) was more and the remaining CaO content was smaller in the higher temperature, and the longer retention time. There is no significant difference in mineral composition depending on the atmosphere condition.

It was calculation result of reaction rate constant that frequency factor A was 0.118 in the atmosphere condition of NH_3 0% combustion, and was 0.117 in the atmosphere condition of NH_3 30% mixed combustion, respectively. We considered that the difference in clinker reaction rate in both atmospheres was within the experimental error range, and that there was no influence of atmosphere condition on the clinker formation reaction after calcination.

Table 4. Results of minerals composition in experiment 1
(C3S: $3CaO{\cdot}SiO_2$, C2S: $2CaO{\cdot}SiO_2$, C3A: $3CaO{\cdot}Al_2O_3$, C4AF: $4CaO{\cdot}Al_2O_3{\cdot}Fe_2O_3$)

	Temperature [°C]	Time [min]	Average of minerals composition [wt. %]				
			f-CaO	C3S	C2S	C3A	C4AF
Test I	1350	0	8.01	23.6	46.3	15.5	7.04
		10	2.12	55.1	22.8	15.4	4.61
		30	0.53	61.6	18.9	14.0	4.75
	1450	0	3.03	50.5	26.5	15.3	4.88
		10	0.38	63.8	15.0	14.1	6.35
		30	0.18	64.4	15.3	13.9	5.68
	1550	0	0.48	63.6	15.7	14.1	5.75
		10	0.15	65.8	13.1	13.3	7.10
		30	0.12	64.8	15.6	11.3	7.73
Test II	1350	0	8.75	21.6	48.7	14.2	6.86
		10	2.04	55.3	23.4	14.3	5.17
		30	0.42	62.7	16.3	13.9	6.30
	1450	0	2.88	52.2	26.3	14.2	4.63
		10	0.34	64.3	16.1	13.0	5.93
		30	0.18	63.2	16.9	13.0	6.20
	1550	0	0.46	62.0	17.4	12.9	6.88
		10	0.15	62.1	16.0	13.3	7.83
		30	0.12	63.6	13.6	13.6	8.58

Evaluation of Clinker minerals and Physical Properties of Cement

Table 5 shows the composition and physical properties analysis results of the fired sample. f-CaO was less than 0.2 wt. % under any combustion conditions, and the sintering reaction progressed sufficiently. It was also possible to obtain the same mineral composition based on chemical analysis.

We confirmed that whether only heavy-oil or NH_3 mixed combustion, the cement clinker has the same minerals and compressive strength satisfying JIS standard.

Table 5. Results of clinker minerals and compressive strength of cement in experiment 2

	Mixed NH_3 (cal. base %)	f-CaO (wt. %)	Bogue Formula (wt. %)				Compressive strength (N/mm^2)		
			C3S	C2S	C3A	C4AF	3d.	7d.	28d.
Test III	0	0.19	49.4	30.5	8.2	9.5	26.2	38.7	59.4
Test IV	30	0.18	50.2	29.9	8.3	9.5	26.4	39.3	60.8

Conclusion

We burned the clinker in the atmosphere-controlled electric furnace, calculated the formation rate of the clinker by Arrhenius type solid-state reaction model. And we fired the clinker in the test furnace under only heavy-oil combustion and NH_3 mixed combustion, measured the compressive strength of the cement made from the clinker.

The results that became clear in this study were shown below.

1. The formation rate of clinker didn't depend on the atmosphere conditions but the temperature conditions.
2. There was no difference in mineral composition and compressive strength of clinker whether only heavy-oil or NH_3 mixed combustion.

Acknowledgments

This work was supported by Council for Science, Technology and Innovation (CSTI), Cross-ministerial Strategic Innovation Promotion Program (SIP), "Energy Career" (Funding agency: JST). Furthermore, the authors would like to thank Combustion Engineering Laboratory, Osaka University to lend the test furnace and support for driving it.

References

1. Agency for Natural Resources and Energy (2016), "Japan's Energy White Paper", September
2. Japan Cement Association (2016), "Cement handbook,"
3. Yasuo Arai (1984), "Material chemistry of cement," *Dainippon Books*, pp. 78.
4. Kazuki Inatsu and Seiji Matuo (2003), "Improvement of the Dynamic Model of a Cement Rotary Kiln in Radiative Heat Transfer and Clinkering Process," *J. EICA*, **8[1]**, pp. 43-54.
5. Japanese Industrial Standards (2002), "Method of X-ray fluorescence analysis of cement," JIS R 5204:2002

6. Japan Cement Association (1997), "Method of quantitative calcium oxide determination," JCAS I-01
7. Japanese Industrial Standards (2015), "Physical testing methods for cement," JIS R 5201:2015

540683 Ammonia-to-Hydrogen System for Fcev Refuelling

Abstract:

Ammonia can play a significant role in fuelling the worldâ€™s growing fuel cell electric vehicle (FCEV) fleet through technologies which allow the decomposition of NH_3, and subsequent extraction and purification of H_2. CSIRO has recently demonstrated a pilot-scale ammonia-to-hydrogen system, incorporating an ammonia decomposition stage with a subsequent membrane-based hydrogen purification stage, at a rate of several kilograms of H_2 per day. Through partnerships with an industrial gas producer and two FCEV manufacturers, the resulting H_2 has been compressed and dispensed into FCEVs. System design, materials, performance and strategies for scale-up and demonstration will be discussed.

540511 My Demonstration Renewable Energy System

Abstract:

The presentation will address the hydrogen and ammonia needed as a fuel and fertilizer on my 320 acre Iowa farm, and the sizing of the resulting demonstration system

The volumetric efficiency of using ammonia as a hydrogen carrier, as contrasted with gaseous hydrogen will be presented.

The chronological development of the system will be reviewed and estimates of the energy need to make the fuel and fertilizer will be addressed.

530425 Chemical Kinetic Modelling of Ammonia-Hydrogen-Air Premixed Flames

Abstract:

Ammonia (NH_3) has been proposed as an alternative fuel for a range of applications, such as internal combustion engines, gas turbines and space propulsion systems [1,2]. Its production is cost-effective and can be achieved through a number of methods, including from biomass and waste [3]. As a flammable substance that does not contain carbon in its composition, its combustion does not emit carbon dioxide (CO_2), which is the main driver of global warming. Ammonia is also considered a safe hydrogen (H_2) carrier due to its low reactivity and can be decomposed onsite into H_2 through catalytic processes [4]. However, NH_3 presents very low flame speeds [5–7], being rather problematic to ignite and burn stably. Additionally, NH_3 produces very high amounts of nitrogen oxides (NO_x), especially under lean conditions, which are involved in atmospheric pollution, causing respiratory problems and acid rain. One of the solutions proposed to enable its use is co-firing ammonia with H_2 in order to improve flammability, and thereby generate more stable flames with higher flame speed [8,9].

To develop systems suitable for using NH_3 and its blends with H_2 in a clean and efficient way, it is necessary to understand its combustion chemistry, which can be achieved through fundamental chemical kinetics studies. The present work uses recently developed kinetic mechanisms [7,10,11] in order to examine critical features associated with the combustion of NH_3 and its blends with H_2 such as ignition delay times and laminar flame speeds. The simulations are compared with available experimental data. Ignition delay time simulations were carried out for NH_3/O_2 mixtures under zero-dimensional shock tube conditions, with 98-99% dilution in argon, under pressure of 1.4, 11 and 30 atm, and temperature ranging from 1564 K to 2489 K. Flame speed and NO_x formation were computed for ammonia/hydrogen/air flames, with ammonia/hydrogen proportion ranging from 0 to 100%, under one-dimensional freely propagating flame conditions, with premixed reactants injected at 298 K, pressures of 1, 3 and 5 bar, and fuel/air equivalence ratios ranging from 0.5 to 1.5. All simulations were done using the Cantera package [12].

The mechanisms of Mathieu and Petersen [7], Otomo et al. [10] and Okafor et al. [11] contain 34, 32 and 25 chemical species, and 159, 213, and 100 reactions, respectively. These mechanisms were reduced to contain 22, 24 and 21 species and 66, 72 and 51 reactions, respectively, in order to reduce the computational demand. The reduced mechanisms based on Mathieu and Petersen [7], Otomo et al. [10] and Okafor et al. [11] are referred to as the MP-R66, the Ot-R72 and the Ok-R51 mechanisms, respectively. Figure 1 shows the predicted ignition delay time of $NH_3/O_2/Ar$ mixtures under pressure of 11 atm and a range of temperature. The mechanisms of Otomo et al. [10] and Mathieu and Petersen [7] show better agreement with experimental data for the ignition delay times, which range from 20 μs at 2500 K to 2500 μs at 1500 K. The mechanism of Okafor [11] predicts higher values, reaching 100 μs at 2500

K and 5000 μs at 1500 K. The reduced mechanisms present slightly higher values than the original ones, reaching 30 μs at 2500 K and 4000 μs at 1500 K for Ot-72.

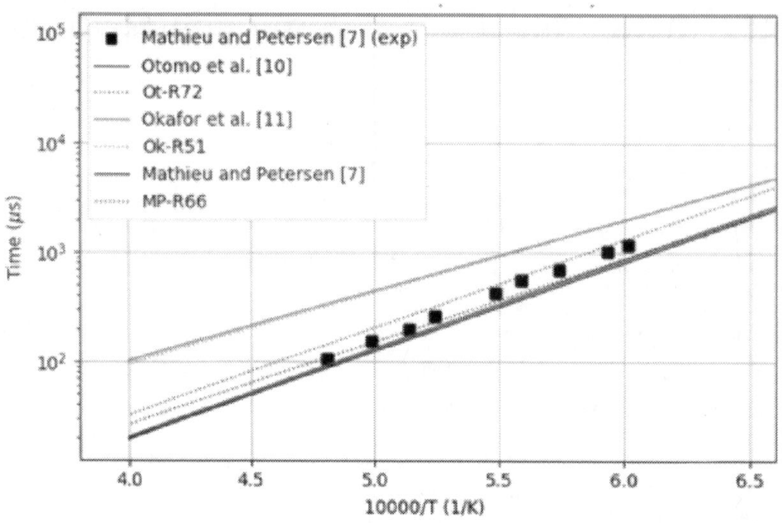

Figure 1. Shock tube ignition delay times for NH_3/O_2 mixtures with 99% dilution in argon at 11 atm for stoichiometric conditions.

The mechanism of Mathieu and Petersen [7] predicts the ignition delay time of $NH_3/O_2/Ar$ mixtures in better agreement with the experiments than the other mechanisms do; however, this mechanism predicts generally lower flame speeds in premixed NH_3/air flames at 1 bar and 298 K, Figure 2. On the contrary, the mechanism of Okafor et al. [11] and its reduced mechanism Ok-R51 over-predict the ignition delay time and the flame speeds, Figures 1 and 2. The mechanism of Otomo et al. [10] and its reduced one Ot-R72 predict flame speeds and ignition delay time in between those from the other mechanisms. In general, the reduced mechanisms predicts similar results as the original ones do, except that the Ot-72 mechanism predicts a lower flame speeds than those predicted by the original mechanism in the lean and rich mixtures, Figure 2. It is worth mentioning that all mechanisms predict peak laminar flame speed at slightly rich mixtures, around equivalence ratio of 1.1. This is comparable with the flame methane/air flames that have a peak flame speed at equivalence ratio of 1.06.

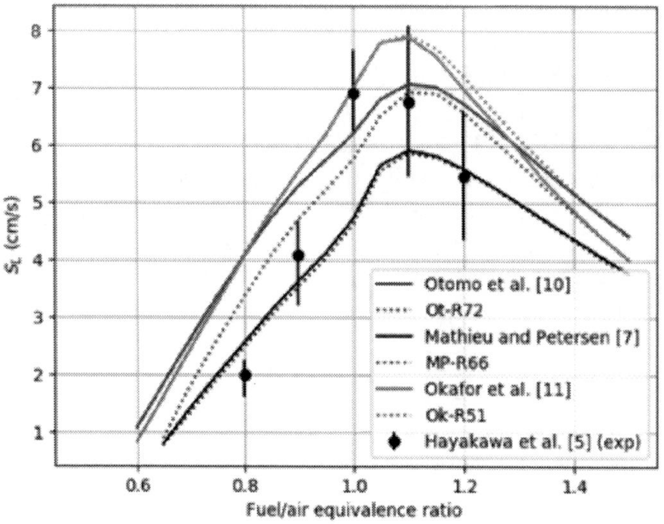

Figure 2. Laminar flame speed of premixed NH_3/air flames at 1 bar and 298 K.

For NH_3 and H_2 blends (Fig. 3) at stoichiometric condition all mechanisms predict lower flame speeds than the measured ones for lower H_2 content. The mechanism of Okafor et al. [11] predicts higher flame speeds for low H_2 content, while the mechanism of Mathieu and Petersen [7] shows the poorest agreement with the experiments for mixtures with low H_2 content. The mechanism of Otomo et al. [10] predicts flame speeds in between those predicted by the other mechanisms. The reduced mechanisms replicate the results of their respective original ones fairly well, Fig.3. It should be pointed out that the laminar flame speed of stoichiometric NH_3/H_2/air flames increase approximately exponentially with the mole fraction of hydrogen in the fuel blends. Thus, it is an effective approach to improve the flame speeds of ammonia flames by blending hydrogen in the fuel mixtures.

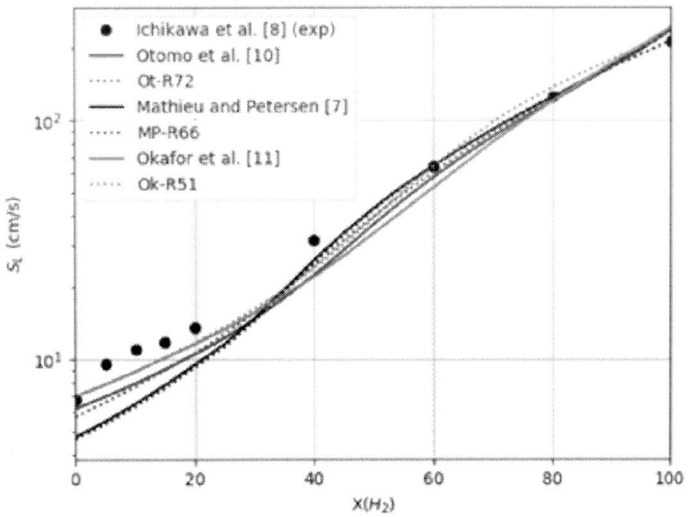

Figure 3. Laminar flame speed for stoichiometric premixed NH_3/H_2/air flames at 1 bar and 298 K, as a function of the mole fraction of hydrogen in the ammonia/hydrogen fuel blends.

All mechanisms predict very high NOx emissions for pure ammonia/air combustion (Fig. 4). The reduced mechanisms present similar behaviour, except for Ot-72. The predicted peak NOx is in lean mixtures, with equivalence ratio about 0.9. The predicted peak NOx mole fraction varies from 2800 ppm to 4600 ppm, indicating a huge scatter of predictions from different mechanisms. For ammonia-hydrogen blends (Fig. 5) the predicted NOx increases with the content of hydrogen in the mixture, owing to the elevated temperature of the mixture, until the mole fraction of hydrogen reaches 80% in the ammonia/hydrogen blends. Further increase in the hydrogen content results in a decrease in NOx production, owing to the decreased content of ammonia in the blends, Fig. 5. The reduced mechanisms predict similar NOx results as those from their original mechanisms, except for Ok-51 at conditions of high hydrogen content in the blends. It should be pointed out that no experimental data available in the literature for NOx in ammonia flames.

In summary, the results shown above clearly indicate that the neat ammonia flames have rather low flame speed, and blending hydrogen with ammonia is an effective approach for improving the flame speed of ammonia flames. However, NOx emission is significantly higher in ammonia/hydrogen flames. The present study shows that the presently available chemical kinetic mechanisms in the literature predicts rather scattered ignition delay time, laminar flame speed, and NOx emissions in ammonia flames.

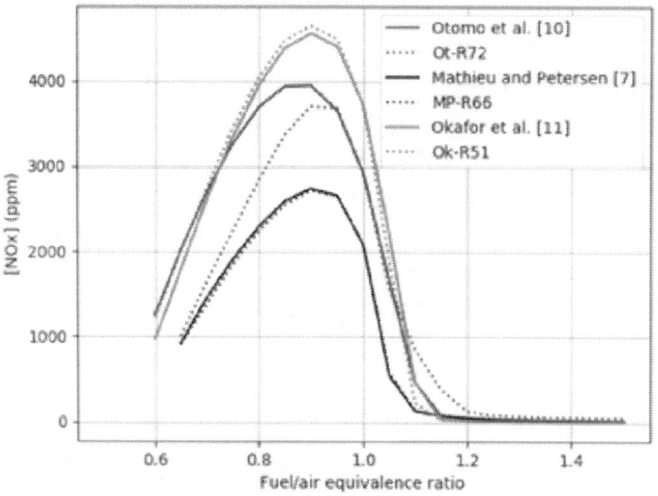

Figure 4. NO_x concentrations for premixed NH_3/air flames at 1 bar and 298 K.

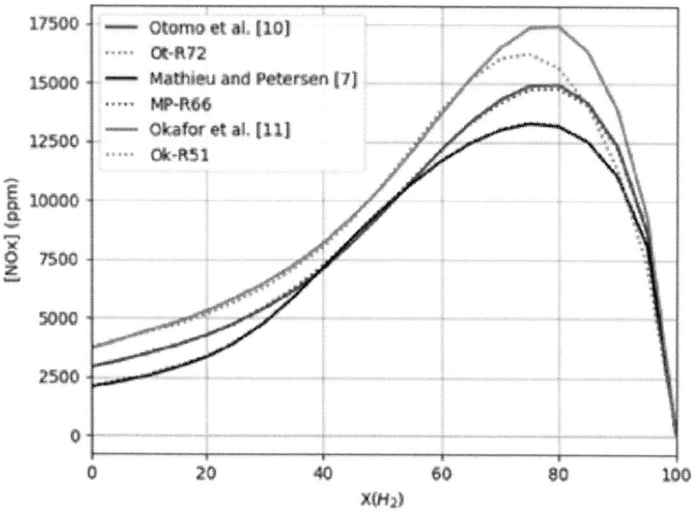

Figure 5. NO_x concentrations for stoichiometric premixed NH_3 + H_2/air flames at 1 bar and 298 K.

References

[1] Zamfirescu C, Dincer I. Ammonia as a green fuel and hydrogen source for vehicular applications. Fuel Process Technol 2009;90:729–37.

[2] Valera-Medina A, Morris S, Runyon J, Pugh DG, Marsh R, Beasley P, et al. Ammonia, Methane and Hydrogen for Gas Turbines. Energy Procedia 2015;75:118–23.

[3] Bicer Y, Dincer I, Zamfirescu C, Vezina G, Raso F. Comparative life cycle assessment of various ammonia production methods. J Clean Prod 2016;135:1379–95.

[4] Mukherjee S, Devaguptapu S V., Sviripa A, Lund CRF, Wu G. Low-temperature ammonia decomposition catalysts for hydrogen generation. Appl Catal B Environ 2018;226:162–81.

[5] Hayakawa A, Goto T, Mimoto R, Arakawa Y, Kudo T, Kobayashi H. Laminar burning velocity and Markstein length of ammonia/air premixed flames at various pressures. Fuel 2015;159:98–106.

[6] Brackmann C, Alekseev VA, Zhou B, Nordström E, Bengtsson PE, Li Z, et al. Structure of premixed ammonia + air flames at atmospheric pressure: Laser diagnostics and kinetic modeling. Combust Flame 2016;163:370–81.

[7] Mathieu O, Petersen EL. Experimental and modeling study on the high-temperature oxidation of Ammonia and related NOx chemistry. Combust Flame 2015;162:554–70.

[8] Ichikawa A, Hayakawa A, Kitagawa Y, Kunkuma Amila Somarathne KD, Kudo T, Kobayashi H. Laminar burning velocity and Markstein length of ammonia/hydrogen/air premixed flames at elevated pressures. Int J Hydrogen Energy 2015;40:9570–8.

[9] Li J, Huang H, Kobayashi N, Wang C, Yuan H. Numerical study on laminar burning velocity and ignition delay time of ammonia flame with hydrogen addition. Energy 2017;126:796–809.

[10] Otomo J, Koshi M, Mitsumori T, Iwasaki H, Yamada K. Chemical kinetic modeling of ammonia oxidation with improved reaction mechanism for ammonia/air and ammonia/hydrogen/air combustion. Int J Hydrogen Energy 2018:1–11.

[11] Okafor EC, Naito Y, Colson S, Ichikawa A, Kudo T, Hayakawa A, et al. Experimental and numerical study of the laminar burning velocity of CH4–NH3–air premixed flames. Combust Flame 2018;187:185–98.

[12] Goodwin DG, Moffat HK, Speth RL. Cantera: An Object-oriented Software Toolkit for Chemical Kinetics, Thermodynamics, and Transport Processes 2017.

AUTHOR INDEX

AUTHOR INDEX